MEASURING THE
Universe

KITTY FERGUSON

MEASURING THE Uni

Walker & Company New York

Our Historic Quest
to Chart the Horizons of
Space and Time

To my brother, David,

WHO, AS A CHILD, MADE HIMSELF ILL
WORRYING ABOUT THE SIZE OF THE UNIVERSE

Copyright © 1999 by Kitty Ferguson

First published in the United States of America in 1999 by Walker Publishing Company, Inc.; first paperback edition published in 2000.

Published simultaneously in Canada by Fitzhenry and Whiteside, Markham, Ontario L3R 4T8

Library of Congress Cataloging-in-Publication Data
Ferguson, Kitty.
Measuring the universe : our historic quest to chart the horizons of space and time / Kitty Ferguson.
p.   cm.
Includes bibliographical references and index.
ISBN 0-8027-1351-3
1. Cosmology.   2. Cosmological distances.   3. Mensuration.
I. Title.
QB981.F346   1999
523.1—dc21   99-19476
CIP

ISBN 0-8027-7592-6 (paperback)

Book design by Ralph L. Fowler

Printed in the United States of America

2   4   6   8   10   9   7   5   3   1

CONTENTS

Measure for Pleasure

❝No science attains maturity until it acquires methods of
measurement.❞
—**Logan Clendening**

There is no task so simple, yet so profound in its consequences, as
the act of taking a measurement. The scientific and technological
foundations of modern society depend upon it. Without it, that
which we call knowledge would have little objective meaning and
our understanding of the natural world would reduce to mythologi-
cal proclamations.

Just look at any attempt to "measure" our world before the
methods of science were developed. At one time or another people
believed that all stars were the same distance from Earth, each em-
bedded on the inside surface of an upside-down bowl; that Earth
was a flat circle in the center of the universe; that Earth and the
universe were just a few thousand years old; that Earth was held up
on the back of a strong fellow; or that Earth was held up on the
backs of elephants, who themselves were standing on the back of a
super-sized tortoise, who himself was resting upon a snake, who
was eating its own tail. Indeed, fantasy-ridden knowledge makes
poor science but fascinating anthropology.

So important is measurement to civilized and rational human behavior that in modern times one might divide our knowledge of the universe into two parts: opinion, and what we know by measuring things. Wherever measurement is not possible, either in practice or in principle, arguments are sure to thrive. Look at politics, religion, and art. These perennial hotbeds of debate and disagreement in society are rarely associated with rational discourse because they are rich in personal viewpoints that are irreconcilable. If you could actually measure how much better one political candidate was than another, or which religion was better than another, all arguments would go away and there would be only one political party and one religion. Scientists argue too, but only on the frontier of discovery where measurements are either nonexistent or unreliable. This is why there are rival religions and rival political philosophies in the world—sometimes leading to war and bloodshed—but there's just one science. Think about it. Nearly every argument you ever overheard (or started) either began or dragged on for want of a measurement. Without systems of measurement, modern life would be difficult but modern science would be impossible.

Often great leaps of knowledge come not from a measurement but from a comparison of one measurement with another. In the late 1500s, when Galileo tested the millennia-old Aristotelian concept that things in motion tend to come to rest, he noted in his experiments that the slipperier the track, the farther along an object slides when pushed. Upon noticing this trend, Galileo correctly surmised that a perfectly slippery track would not impede the motion of a sliding object at all, enabling it to continue in motion forever. A century later, Isaac Newton would follow with a fundamental and universal law: Objects in motion tend to stay in motion unless acted upon by an outside force. When combined with his universal law of gravitation, Newton laid a scientific foundation that led directly to the industrial revolution, the measurement of our solar system, and ultimately the era of space travel. Our ability to explore the planets with directed space probes may be a modern analogue to the fifteenth- and sixteenth-century voyagers who used their knowledge (limited though it was) of Earth's shape, size, and land patterns to plot their round-the-world journeys. Columbus knew the Earth was round just as Newton knew the universe was large.

And at the end of the nineteenth century, the scientists Michelson and Morley measured the speed of light more accurately than ever before. They measured it in the same direction as Earth's motion around the sun. They measured it in the opposite direction of Earth's motion around the Sun. They measured it sideways to Earth's motion around the Sun. They were trying to see how Earth's velocity through space would either add or subtract from the velocity they measured for light. It didn't. In what was perhaps the most famous nonresult in the history of physics, Michelson and Morley determined that the speed of light was constant no matter how you measured it. This became the experimental foundation for Einstein's theory of relativity, which heralded the era of modern physics.

Measurement itself is not a free pass to scientific enlightenment. In all cases, accuracy matters more than precision. For example, the impressive precision of digital clocks has given society a false sense of accuracy. Several years ago, I was one of the dozens of thousands of people who assembled in Times Square, New York City, to celebrate New Year's Eve. While there, I took note of three different digital clocks that were part of large neon billboards. They each boasted high precision: Two of them displayed time to the tenth of a second; the other displayed time to the hundredth of a second. A fourth digital clock was high on a building in the middle of Times Square, which was the clock that most people watched. At midnight, whenever that was, *none* of the clocks agreed. A full ten seconds separated the fastest from the slowest clock. At least three clocks, and quite possibly all four, were wrong. Most people assembled for the occasion took no notice, but I was devastated. As the adage goes, "The person with one clock knows the time. The person with two clocks isn't sure."

The rotating Earth served human time-keeping needs quite well for millennia. But when we keep time with atomic clocks, which are extremely accurate and of extremely high precision, we find that Earth's rotation rate is not absolutely reliable. A simple way to measure this sad fact is to watch a star cross your meridian and see how much time elapses until it does it again. This will be Earth's period of rotation. Careful measurements of this "sidereal" period have shown that Earth's rotation is slowing down from the combined tidal forces of the Moon and Sun. The sloshing of the oceans back and forth on the continental shelves creates a drag on our day. Every couple of years or so we must introduce a leap-second into the calendar to compensate. In gravitational response to Earth's slowdown, the Moon is spiraling away from us at a rate of about a centimeter per year, slowly increasing the length of the month. Yes, days and months used to be shorter and the Moon used to be closer (and therefore bigger) in the sky. Behold the power of measurement.

When astrophysicists are at work, chances are we are busy trying to measure something about the universe. How tall? How far? How luminous? How massive? How hot? How big? How fast? How energetic? No scale is off-limits: We might be measuring the distance to the Moon with a laser pulse bounced from a mirror left there by the Apollo astronauts, or we might be measuring the distance to the edge of the observable universe using data gathered from the largest telescopes in the world. Modern bookshelves are filled with stories of cosmic discovery—from the search for planets to the search for black holes to the search for a theory of everything. Occasionally, however, an author comes along who dares to describe how science works, who dares to find its underbelly and remind us

that the romance and pleasure of cosmic discovery lies not necessarily in your experimental results but in the journey of measurements
that led to them.

That author is Kitty Ferguson, a chamber musician turned science writer, who is already distinguished as one who can explain
complex things—from the life and times of cosmic objects (in *Prisons of Light: Black Holes*) to the life and times of cosmic physicists
(*Stephen Hawking: Quest for a Theory of Everything*).

—NEIL DEGRASSE TYSON

ACKNOWLEDGMENTS

I want to thank the following people, who have helped make this book possible by reading all or parts of it, discussing their work with me, answering my questions, supplying background material and information, making suggestions and corrections, and coming up with ways to make the explanations clearer: Judy Anderson, Boyd Edwards, Caitlin Ferguson, Yale Ferguson, Wendy Freedman, Carlos Frenk, Margaret Geller, Owen Gingerich, Walter Gorzegno, Stephen Hawking, Howard Helms, Jill Knapp, Helen Langhorne, P. Susie Maloney, Robert Naeye, Bruce Partridge, Saul Perlmutter, Barbara Quinn, Allan Sandage, Bill Sheehan, Patrick Thaddeus, and David Vetter.

MEASURING THE

Universe

Tilting with Windmills

{1951}

When I was nine years old, my father suggested one morning that
he, my brother, and I go out and measure the height of the windmill
on my grandparents' farm. David and I agreed that was a fine idea.

How would we do it? I wondered. Climb the windmill, of
course . . . at least Dad would since David and I wouldn't be allowed
to try anything so dangerous. When Dad reached the top, there'd
still be a problem, because we didn't have a measuring tape that was
long enough. Would he take a yardstick and mark off the yards on
the windmill as he climbed? Or maybe he would drop the end of a
long rope from up there and cut it off, and when he climbed down
we'd measure the cut-off piece. That must be the plan, since Dad
had said my brother and I would help him.

David suggested that Dad wouldn't need to climb the windmill
at all. We could throw something over the top, just clearing it.
Yes—I interrupted—attach a rope to the thing we threw, a rope

with inches and feet marked on it, and then pull back on it gently so it would catch on the top of the windmill, and we'd see what the measurement was to the ground! No, no, cried David, who was two years younger than I but already very mathematically minded. We would measure the curve the object followed through the air. Good thinking, said my father, but, practically speaking, this solution was more complicated than the original problem.

I asked whether we might walk away from the windmill and measure how much smaller it looked as we got farther away. More good thinking, said Dad, but there was a better way.

He gave us a hint: On a sunny day, no one would have to climb the windmill or take a walk or risk wrecking the windmill with a bad throw . . . and the only tools we'd need would be a yardstick and our eyes and brains and a pencil and paper to do some calculations. He also said that although at this latitude it would be possible to measure the windmill precisely at noon, it would be easier at another time of day.

Neither my brother nor I was clever enough to see where this was leading until Dad said, "The windmill does more than just pump water, you know. It casts a shadow, and so does a yardstick," and then David and I began to understand how the trick could be done. We would stand the yardstick upright and measure its shadow. Then we would measure the windmill's shadow. If a shadow *this* long went with a three-foot stick, then a shadow THIS long went with a windmill of thus-and-so height. David and I didn't know how to make the comparison. Dad taught us how and then told us there was actually a more primitive way we could find the answer: Wait for the time of day when the yardstick cast a three-foot shadow. At that time the length of the windmill's shadow would be the same as the height of the windmill. David and I considered that but decided to use our newly acquired math skills first. We checked our answers by sitting out in the Texas sun, watching the two shadows creep along the ground.

That's how we measured the windmill, while above our heads the giant structure thrummed and creaked with the watery, metallic sounds windmills in central Texas made in those days, doing its work, turning and pumping, adjusting its angle to catch a stronger breeze, oblivious to the children below who had captured its shadow.

I was elated. It seemed we had outwitted the windmill without so much as touching it. We didn't ask why we were doing this. There was no need whatsoever for us to know the height of a windmill that wasn't even our own windmill. Instead, we reveled in our wondrous secret: not the height of the windmill but the knowledge of how to find it out.

Measuring is one of the more practical uses of mathematics, but the ability and desire to measure aren't always wrapped up with the need to know useful answers. Letting numbers take us where we can't go in person—whether that's to the top of a windmill or to the origin and borders of the universe—has been and still is one of humankind's favorite intellectual adventures. At the end of the twentieth century, our success has outrun practical requirements by billions of light-years.

Compared with the adventure of discovering them, the statistics themselves often seem terribly dry: The Sun is 149.5 million kilometers away (mean distance). The nearest star is 4.3 light-years. The "Local Group" of galaxies covers an area about 3 million light-years in diameter. The distance of the edge of the observable universe is 8 to 15 billion light-years. The largeness of these numbers renders them meaningless. You remember them for a day or two or maybe long enough for a school exam . . . and then forget them. Science trivia.

But these measurements aren't trivial at all when you realize how hard-won they are and what ingenuity it took and still takes to find them out. Imagine what it would be like if no one knew them. Suppose you and I still wondered whether or not all of the pinpricks of light in the night sky are the same distance from us. Suppose none of our contemporaries could tell us whether the Sun orbits the Earth, or vice versa, or even how large the Earth is. Suppose no one had guessed there are mathematical laws underlying the motion of the heavens? How would . . . how did . . . anyone begin to discover these numbers and these relationships without leaving the Earth? What made anyone even think it was possible to find out "how far" without going there. Without climbing the windmill.

In *Measuring the Universe*, we'll forget we know the measurements or how to make them, and join our ancestors as they tease this information out of a sky full of stars. The laboratory isn't a neat, sterile room where carefully controlled experiments take place. Events in the heavens happen in their own good time and not before, and they are often not repeatable. Astronomers have learned to take what's on offer and make the best of it.

Humanity's point of view is sorely limited. Until recently, the Earth was the only platform from which to take measurements. In the late 1960s and early 1970s, astronauts traveled to the Moon and looked back at our planet from space. Unmanned space missions have probed the far reaches of the solar system. But by universal standards, by the standards of the distances human beings have learned to measure and still hope to measure, how pitifully close to home that is!

This book is a chronicle of how men and women over the course of two and a half millennia have built a ladder of measurement from their doorsteps to the borders of the universe, and how the adventure has changed ideas about the shape and nature of the universe and our place in it. It is not a history of all astronomy. There have been important discoveries and theories, both in Western astronomy and in that of other cultures, that have had no direct bearing on our present knowledge of distances, size, and shape. With regret, I have left them out, though the temptation to embark on long digressions from the main theme of the book has been almost irresistible.

Measuring the Universe broadens our focus in another direction, however, to examine the context in which these discoveries occurred, for this is a story inextricably bound up with social, political, and intellectual history. Why did a particular discovery or measurement happen when and where it did? What was it about that time and place—that society, that mind-set or intellectual milieu, the available technology, the chain of previous discovery, the way some random occurrences fell out—or what was it about a specific individual—that precipitated this advance in knowledge?

The story of wanting to know "how far"—to make ridiculously out-of-reach measurements—must surely have begun before the era of recorded history, but the known story of humankind's success began some 2,200 years ago in north Africa near the mouth of the Nile with the measurement of the circumference of the Earth, long before anyone was able to circumnavigate it. The Hellenistic librarian Eratosthenes didn't need to know the circumference of the Earth. Nevertheless, he set about to measure it, and he did it in a surprisingly simple way. Eratosthenes is called the father of the science of Earth measurement, geodesy, which fittingly rhymes with *odyssey*.

His remarkable achievement exemplifies the lesson David and I learned from our father: What can't be measured directly—what it is unthinkable that anyone should ever measure directly—*can* be measured in roundabout, inventive ways. Today scientists are attempting to determine the distance to objects near the borders of the observable universe, far beyond anything that can be seen with the naked eye in the night sky, and to measure time back to the origin of the universe. The numbers are indeed too enormous for our minds to comprehend. Nevertheless, generations of minds have figured them out, one resourceful step at a time, each step building upon the last. Technology has improved astoundingly, particularly in the twentieth century, but progress has owed as much to raw ingenuity. It still does. Frontier inventiveness is not out of date.

With the benefit of hindsight, you and I may be tempted to exclaim, "Of course! Why . . . I could have thought of that!" Many of the methods for measuring distances to out-of-reach places are simple enough for nearly everyone to understand—only a little more complicated than measuring my grandfather's windmill. But, to have discovered these methods for the first time . . . how impossibly clever!

A Sphere
with a View

{400–100 B.C.}

> The great mind, like the small, experiments with different alternatives, works out their consequences for some distance, and thereupon guesses (much like a chess player) that one move will generate richer possibilities than the rest. . . . It still remains to ask how the great mind comes to guess better than another, and to make leaps that turn out to lead further and deeper than yours or mine. We do not know.
>
> **Jacob Bronowski**

Eratosthenes of Cyrene, who held the distinguished title of director of the Alexandria Library from 235 to 195 B.C., also had two nicknames: "Pentathlos" and "Beta." Pentathlos was a name for athletes who entered the pentathlon, which required five skills. Eratosthenes was not an athlete. The nickname implied he was a jack-of-all-trades. The word *beta* stood for the letter B or number two or second. Put those together and you get "jack-of-all-trades and master of none." Whether Eratosthenes's colleagues gave him those names out of fondness or scorn isn't known, but whatever mockery this venerable polymath may have endured, Beta is remembered while most who dubbed him that have long since been forgotten.

Eratosthenes's accomplishments were, indeed, numerous and eclectic. He attempted to fix the dates of the major literary and political events since the conquest of Troy; he composed a treatise about theaters and theatrical apparatus and the works of the best-known comic poets of the "old comedy"; he suggested a way of solving a problem that had tantalized mathematicians for two centuries—"duplicating a cube"; and he let his voice be heard on the subject of moral philosophy and felt it essential to criticize those who were "popularizing" philosophy . . . "dressing it up in the gaudy apparel of loose women." But it was none of those achievements that won him his place in the history books. What modern school-children learn about him is that he invented "the sieve of Eratosthenes"—a method for sifting through all the numbers to find which are prime numbers—and that he discovered a way to measure the circumference of the Earth with astounding accuracy.

The idea that scholars before Columbus believed the world was flat is a fable created in modern times. Admittedly, the shape of the Earth probably wasn't of much daily practical interest to most people in the ancient world. However, already long before Eratosthenes, those few who were wondering about it at all were not seriously suggesting that the Earth was any shape but spherical. The Pythagoreans, a school of thinkers revered for their genius in mathematics and music, had decided as early as the sixth and fifth centuries B.C. that the Earth is a sphere. A century before Eratosthenes, Plato pictured a cosmos made up of spheres within spheres, nested one within the other, with a spherical Earth at the center. Only a little later than Plato, Aristotle vigorously subscribed to the idea of a spherical Earth, and his defense proved convincing not only to the ancient world but also to the Middle Ages.

Aristotle rested his case partly on observational evidence: During an eclipse of the Moon, the shadow cast by the Earth on the Moon is always curved. Also, in Aristotle's words:

> *There is much change, I mean, in the stars which are overhead, and the stars seen are different, as one moves northward or southward. Indeed there are some stars seen in Egypt and in the neighborhood of Cyprus which are not seen in the northerly regions; and stars which in the north are never beyond the range of observation, in those regions rise and set. All of which goes to show not only that the Earth is circular in shape, but also that it is a sphere of no great size: for otherwise the effect of so slight a change of place would not be so quickly apparent.*

Aristotle speculated that the oceans of the extreme west and the extreme east of the known world might be "one," and he reported with some sympathy the arguments of those who had noticed that elephants appeared in regions to the extreme east and the extreme west, and who thought therefore that those regions might be "continuous."

Aristotle's philosophy also argued for a spherical Earth. He had concluded that five elements—earth, air, water, fire, and aether—each have a natural place in the universe. The natural place for the element earth is at the center of the universe, and for that reason earth (the element) has a natural tendency to move toward that center, where it must inevitably arrange itself in a symmetrical fashion around the center point, forming a sphere. Aristotle reported that mathematicians had estimated the Earth's circumference as 400,000 stades; that is, about 39,000 miles or 63,000 kilometers (more than half again as large as the modern measurement). No record survives of the method used to arrive at that number.

The Intellectual Spoils of War

When Aristotle died in 322 B.C. at age sixty-two, the military campaigns of his most highly achieving pupil, Alexander the Great, had just ended with Alexander's death. Vastly widened mental horizons were part of Alexander's extraordinary legacy. His campaigns had carried Greek knowledge, language, and culture throughout Asia Minor and Mesopotamia as far east as present-day Afghanistan and Pakistan, all the way to the Indus River, as well as to Palestine and Egypt. The culture of Greece and its colonies and the cultures of the conquered peoples began to mix and enrich one another. This was the dawn of the Hellenistic era, as opposed to the Hellenic. That is, Greekish, as opposed to Greek.

At the time of Alexander's and Aristotle's deaths, within a year of one another, Athens was still the undisputed center of the intellectual world. That preeminence was not to last. Alexander's generals divided his empire, and Ptolemy's portion was Egypt and Palestine. He made Alexandria, near the mouth of the Nile, his capital. This already prospering city began to grow in size and splendor, and Ptolemy and his successors, reputedly ruthless in their exploitation of the lands under their control, amassed a surplus of wealth, some of which they chose to spend on literature, the arts, mathematics, and science. Scholars are divided as to which Ptolemy should get the credit (Ptolemy's successors were also named Ptolemy), but either the first or the second, and perhaps it took both, extended the royal patronage to found a library and a museum. The word museum meant "temple to the muses," both a religious shrine and a center of learning.

Meanwhile the old, justly famous schools across the sea in Athens, schools founded by Plato, Aristotle, Epicurus, and the Stoics, were no longer producing vibrant new ideas to quite the extent they once had, though they were still the places a young man of Eratosthenes's time would have wished to go for his education. Alexandria began to rival and eventually supplanted Athens as the focal point of the thinking world, and the Alexandria Museum and Library became the premier research institution. The library grew large, containing by one ancient estimate nearly 500,000 rolls. The salary of the director, or librarian, came from the royal coffers.

The story (now thought to be apocryphal) is that the contents of the library at Alexandria were burned to heat the public baths for six months in the seventh century A.D. Whether the destruction occurred quickly and calamitously then or, more likely, gradually through neglect and the many political, military, and religious turns of fortune that affected the city of Alexandria, the loss was the symbol and symptom of a greater tragedy: the widespread disappearance of any perception that such intellectual achievement was of value. By the 600s, there was probably little left to burn. It took centuries for humanity in the Western world to reach again an intellectual level on a par with the civilization that had produced that lost collection. But in the third and second centuries B.C., all this was still many centuries in the future. The Alexandria Library was in its heyday.

The scholars connected with this august institution and their forebears in the Hellenic world would have been mystified by the present-day concept of "science" as a distinct category of knowledge and pursuit of knowledge. Some modern words have evolved from terms they used to describe similar areas of interest, but the modern words don't have precisely the same meaning these had in ancient Athens and Alexandria. Some examples: *peri physeos historia* (inquiry having to do with nature); *philosophia* (love of wisdom, philosophy); *theoria* (speculation); and *episteme* (knowledge). Hellenistic scholars thought of "physics" as one of three branches of philosophy. The other branches were "logic" and "ethics."

Another key difference between the ancient and modern ways of thinking is that Hellenic and Hellenistic scholars tended to be somewhat scornful of the notion that their effort might serve mundane, practical purposes. They preferred to think of it as contributing to wisdom, or improvement of one's character, or leading to greater appreciation of the beauty of the universe and understanding of its creator. The life of a scholar, the life of "contemplation," was considered an exquisitely happy life. Doctors, whose efforts were intended to have more everyday practical value, were apt to differentiate themselves entirely from the "philosophers," whose work was its own reward, an end in itself, not a means to an end. The Ptolemys were of course far from displeased when research could be applied to problems connected with weaponry, but their financial support and their efforts to outbid all competitors when it came to collecting the masterpieces of Greek literature and encouraging distinguished scholars to flock to Alexandria were far more strongly motivated by desire for prestige—to add to the luster and apparent power of the dynasty. It seemed not to occur to these men and women in ancient times that scholarly endeavors might hold the key to material progress.

Theirs was also a perspective in which *how* to solve a problem was as interesting as actually solving it, often more so, an attitude arising partly out of necessity, for Greek and Hellenistic scholars were fascinated with questions that they lacked the technology to answer definitively. A modern analogy might be the typical "word problem" in grade school. Let's say you're presented with the question: If you ride your bicycle at an average speed of thirty miles per hour, and it takes you ten minutes to get from home to school, how far is school from home? You do not immediately start quibbling that thirty miles per hour is not an accurate measurement of the speed you normally ride, that it actually takes you twelve minutes to get to school, and that this exercise isn't going to end with anyone knowing how far from home your school really is. No. What everyone is interested in is your showing that you understand how to solve the problem. Move back a step and imagine that it is also up to you to invent the method for solving it—that no one, in fact, has ever even thought it possible to calculate the distance from your home to your school and that you can't ride there to measure it directly. You have put yourself in the shoes of Eratosthenes and others whose work this chapter describes, a situation that allows, indeed encourages, the formation of hypotheses, sometimes out of thin air . . . statements such as, "We don't know that this is true, but let's assume for a moment that it is, and see where that gets us." Or even such a statement as, "We know that this is *not* true, but let's pretend for the moment that it is and ask 'what then?'" To criticize the results of an exercise like that by saying the results are "wrong" (i.e., do not accord with twentieth-century findings) is to miss the point.

The scholarly mind-set of his era partly explains why Eratosthenes took on what might seem to have been an impossible problem and tried to find an answer that was of no use to the ancient world: the circumference of the Earth. His success was rooted in the widened horizons and the mixture of cultures that characterized the Hellenistic world. It also had a great deal to do with the sort of man he was.

A Sunlit Well at Syene

Eratosthenes, "son of Aglaos," was born not in Egypt or Greece but in the ancient city of Cyrene, on the northern coast of Africa in what is now Libya. Citizens of Crete and Santorini had founded Cyrene some 350 years earlier, and it had become one of the most cultured cities of the Hellenistic world, though still subordinate to the Egypt of the Ptolemys. Cyrene counted some distinguished figures among its citizenry. The founder of the Cyrenaic school, Aristippos, had been a pupil of Socrates. Aristippos's daughter, Arete, followed him as head of the school, and her son, Aristippos II,

succeeded her. He was nicknamed "Metrodidactos," which trans-
lates as "mother-taught."

The date given for Eratosthenes's birth is the "126th Olympiad,"
referring to the Olympic games that took place every four years. In
modern dating, that puts it between 276 and 273 B.C. He received
most of his education in Athens at the feet of eminent scholars of
the New Academy and the Lyceum. Plato and Aristotle had originally founded these schools (in Plato's day the New Academy had
been simply the Academy), and though much had changed about
them by the time Eratosthenes arrived, they were still the most
prestigious educational institutions.

By the middle of the century, Eratosthenes had written a few
philosophical and literary works, and some of these had come to the
attention of Ptolemy III Euergetes. The "brain-drain" from Athens
being in the general direction of Alexandria, Eratosthenes in about
244 B.C. agreed to move there and become a fellow of the museum
and tutor to the prince, Philopator. It is perhaps not to Eratosthenes's credit that his pupil, though a patron of arts and learning,
gained a reputation for dissipation and crimes that rivaled Nero's
and Caligula's later in Rome.

In the course of time, Eratosthenes became a senior (alpha)
fellow of the museum, and upon the death of the chief librarian he
took over that post—an unparalleled vantage point from which to
keep up with everything that was going on in the scholarly world.

Unfortunately, none of Eratosthenes's many works have survived except in fragments. It's not even certain that all the fragments
are genuine. Most information about him comes from reports and
references of others. However, there is enough to tell that Eratosthenes's measurement of the Earth and his motive for attempting it
were rooted in his eclectic and far-ranging knowledge and interests.
Eratosthenes was a man of the world, in the literal sense of those
words. He refused to categorize people as Greeks opposed to barbarians, adopting a new Hellenistic global point of view that had

begun to replace the more parochial mind-set of Greece in earlier centuries. He collected information about the people, products, and geography of far-flung areas. He wrote about the history of geographic measurement, recalling old ideas going back to Homer about the size, shape, and geographic layout of the Earth. In fact, he did nothing less than pull together virtually all the geographic knowledge that had been accumulating up to his own time.

Over the centuries, this material had taken a variety of forms. It came from traders, explorers, travelers—as well as mathematicians and philosophers—and it ranged from fantastic tales to more straightforward reporting, from speculation to measurements and estimates resting on what were probably recognized as shaky assumptions. Among the more reliable sources were eyewitness accounts of Alexander the Great's expeditions and the measurements and records of distances covered on those marches. There were itineraries of coastal voyages and maps and charts connected with them. There was a treatise on harbors by Timosthenes, the admiral of the Ptolemaic fleet, who also studied the winds. There was a book titled *On the Ocean* by the merchant sea captain Pytheas, who in about 320 B.C. sailed north along the coast of Spain and France and reached Cornwall, then continued all the way up to the Orkneys and the Shetlands to latitudes near those of the midnight sun. Pytheas took bearings throughout his voyage and recorded them in his book, which also had descriptive passages: "The barbarians showed us where the Sun keeps watch at night, for around these parts the night is exceedingly short, sometimes two and sometimes three hours, so that only a short interval passes after the Sun sets before it rises once more." Eratosthenes respected Pytheas's information, though many other scholars were contemptuous and disbelieving. Living as Eratosthenes did in Hellenistic Egypt, he may also have known of centuries-old and astoundingly accurate Egyptian geographic calculations.

Eratosthenes's expertise on longitude and latitude surpassed any other of his day or earlier. His predecessors had divided the map into zones. He took that work several steps farther by improving on a map devised about twenty-five years before his birth by a man named Dicaearchus of Messene, who had divided the known world by using two lines or bands that intersected one another—one running east-west, the other north-south. On Eratosthenes's revised map the two lines crossed at Rhodes, a little to the east of where Dicaearchus's lines had met. The horizontal line passed near Gibraltar (then known as the Pillars of Hercules), ran the length of the Mediterranean, and then followed the Taurus chain of mountains in southern Turkey. (Toros Daglari on later maps.) The path of that line is remarkably close to the course of the thirty-seventh parallel—an impressive achievement without the benefit of the mathematical and astronomical knowledge that would go into later mapmaking. It was not yet possible to calculate latitude with very great precision and virtually impossible to determine longitude (which would prove to be a problem in Eratosthenes's measurement of the Earth). Eratosthenes's vertical line, following the Nile, doesn't line up so perfectly with Rhodes on modern maps. He drew six more vertical lines at intervals between the western and eastern boundaries of the inhabited world, and six more horizontal lines at intervals between its northern and southern boundaries. In addition, he established and measured geographic zones, dividing the world horizontally between the tropical region, the temperate region, and the polar circles.

Eratosthenes was also well acquainted with state-of-the-art geometry, both from Euclid's brilliant summing up about twenty-five years before Eratosthenes's birth and from his association with Archimedes, an extraordinary genius and world-class eccentric. There is the familiar tale of Archimedes' solving a mathematical problem in his bath, leaping from the water, and running naked through the streets shouting "Eureka!" This avid mathematician eventually lost his life when Roman troops sacked Syracuse. Archimedes, so the story goes, was drawing a mathematical figure in the sand when a Roman soldier (who had missed hearing an order from his superiors to respect the person of this famous old man) asked him to pack up and move along. Archimedes unwisely told the soldier not to interrupt his thought process.

The Hellenistic world revered Archimedes as an inventor (though he himself dismissed those practical achievements as unworthy) and a useful man in wartime. According to legend, he destroyed a Roman fleet by using burning mirrors. The Middle Ages respected him as an engineer and a wizard and credited him with the invention of the Staff of Archimedes, a stick with a small flat disk that could be run up and down it, so that an observer holding it up to the Sun and noting the distance from disk to eye could derive the Sun's apparent diameter. Modern history and mathematics books recall Archimedes as a brilliant mathematician and geometer who contributed significantly to the understanding of circles and spheres. Archimedes was in the habit of sharing his discoveries and his methods with Eratosthenes and even dedicated his greatest work, *Method*, to him. Eratosthenes must have welcomed another scholar who was almost as eclectic as he was himself.

Eratosthenes's thoughts stretched to the horizon in all directions. Perhaps it follows that he would have longed to know not only what was beyond those horizons but how far "beyond" was. Mapping and systematizing things geographically was his bent. Would he not have been unusually curious about how large the total map was? How remarkable if it really should turn out to be, as Aristotle speculated, "a sphere of no great size"! Eratosthenes's thoughts often took a historical turn, and he was aware of previous attempts to measure the Earth or estimate that measurement. Would he not have wanted to try his own hand at it, using Euclid's and Archimedes' newer understanding of geometry?

There is still one circumstance to be mentioned—a simple, trivial matter, yet Eratosthenes's successful measurement of the circumference of the Earth would not have taken place without it. A happenstance, perhaps, that such a small gem of information reached the ears of this man who realized what it meant and what could be done with it. It is true that the fact that this snippet of news reached him did have something to do with the broadened mental horizons of the world, with improved communications from remote areas, with Eratosthenes's own world centering on northern Africa, and with his habit of keeping his ears and eyes open and wanting to know everything and anything. He was indeed the right man in the right time and place. Perhaps there was no other so likely to run across this back-page news and recognize its worth:

In a well located at Syene (near modern Aswan), on the day of the summer solstice, a shaft of sunlight penetrated all the way to the bottom of the well.

Eratosthenes knew that this signified that the Sun was shining directly down at Syene, not at an angle, which meant that Syene was on the "Tropic." A stick set up at noon at Syene on the day of the summer solstice would not cast a shadow. A stick set up at Alexandria (which he thought was the same longitude as Syene) *would*. Accordingly, Eratosthenes set up a stick at Alexandria on the day of the summer solstice and measured the angle of its shadow when that shadow was at its shortest.

Figure 1.1 shows the stick at Alexandria and its shadow and what "the angle of the shadow" means. If you draw a straight line from the point marked Alexandria (where the stick is casting a shadow) to the center of the Earth, and a second straight line from the point marked Syene (where the stick casts no shadow) to the center of the Earth, those lines will of course meet at the center of the Earth. The question is: What is the angle created by those two lines where they meet? The answer, as Eratosthenes knew, is that

The angle where the two lines meet at the center of the Earth

The angle of the shadow

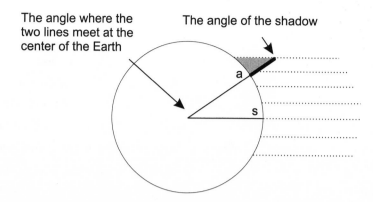

Figure 1.1 Because the Sun's rays are running parallel as they strike the Earth, if a line is drawn from Alexandria (a) where the stick casts a shadow, to the center of the Earth, and a second line from Syene (s) where there is no shadow, to the center of the Earth, the angle where those two lines meet will be the same as the angle of the shadow at Alexandria.

the angle at the center of the Earth and the angle of the shadow at Alexandria will be the same angle. Figure 1.2 illustrates Eratosthenes's measurement.

Eratosthenes found that the shadow angle at Alexandria was $7\frac{1}{5}$ degrees, and so he knew that the angle between the "Syene/Alexandria lines" (meeting at the center of the Earth) was also $7\frac{1}{5}$ degrees. A circle has 360 degrees, and it is a simple process to find out how many of the Syene-Alexandria angles ($7\frac{1}{5}$ degrees) it will take to make 360 degrees. Think of the cross section of the Earth as a pie and the two lines coming from Syene and Alexandria as cutting out a wedge of pie. How many wedges of that size can be cut from the whole pie? Divide 360 by $7\frac{1}{5}$, and it comes out to 50 wedges. Eratosthenes calculated that the distance between Syene and Alexandria at the surface of the Earth—at the pie-crust edge of the pie—is 5,000 stades. (There is a story that he sent a man to pace it off for him.) He multiplied 5,000 by 50 and concluded that the distance all the way around the Earth—the circumference of the Earth—is 250,000 stades. He later fine-tuned this to 252,000 stades.

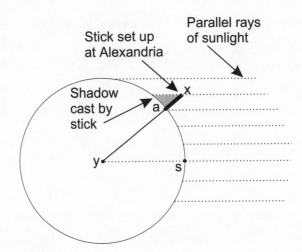

Figure 1.2 Because the sunlight shone all the way to the bottom of the well at Syene (s), Eratosthenes knew that the Sun was shining straight down on the Earth there. He set up a stick at Alexandria (a), where the Sun wasn't shining straight down, and he measured the angle (x) of the shadow cast by the stick. He knew that because the Sun's rays all run parallel as they strike the Earth, the angle (y) where a line drawn straight down from Alexandria and a line drawn straight down from Syene would meet at the center of the Earth would be the same angle as the angle of the shadow cast by the stick (x). If Syene is due south of Alexandria, then the distance between Syene and Alexandria must be the same fraction of the Earth's total circumference as the angle at x or y is of 360 degrees.

What is this odd unit of measurement, the stade? That question brings up a problem in evaluating Eratosthenes's result. Whether or not that result matches modern measurements for the circumference of the Earth depends on the length of stade he was using. If there are 157.5 meters in a stade, Eratosthenes's result comes to 24,608 miles (39,690 kilometers) for the circumference of the Earth. That is very near the modern calculation: 24,857 miles (40,009 kilometers) around the poles and 24,900 miles (40,079 kilometers) around the equator. After he found the circumference, Eratosthenes calculated the diameter of the Earth as 7,850 miles (12,631 kilometers), close to today's mean value of 7,918 miles (12,740 kilometers).

Another way of figuring a stade was as one-eighth or one-tenth of a Roman mile, and that would make Eratosthenes's result too large by modern standards. There was one additional small difficulty. Eratosthenes assumed that Syene lay on the same line of longitude as Alexandria. Actually, it did not.

But this is nit-picking! No apology need be made for Eratosthenes. First of all, he arguably came astonishingly near to matching the modern measurement. Second, he found a way to solve this problem by the imaginative use of geometry. His *method* was ingenious and correct. If the numerical result is a little fuzzy because of a lack of agreement about the length of a stade and the impossibility of determining longitude precisely, that does not detract from the brilliance of his achievement or of the intellectual leap involved in recognizing that measuring the Earth's circumference *could* be done and *how* it could be done.

Eratosthenes's curiosity went beyond the Earth. He also considered the astronomical questions of his day. When it came to measuring the distances to the Sun and the Moon, he must have realized that he had no tool at his fingertips to equal the news about the well in Syene. Nevertheless, he gave it a try, with far less success than he had in measuring the Earth's circumference.

Aristarchus of Samos

Another Hellenistic scholar, Aristarchus of Samos, also tried to measure the distances to the Moon and Sun. Little biographical information exists about him. He lived from about 310 to 230 B.C. and was already a grown man when Eratosthenes was born. The island of Samos was under the rule of the Ptolemys during Aristarchus's lifetime, and it is possible that he worked in Alexandria. Archimedes was certainly aware of his contributions, and Aristarchus knew of Eratosthenes's measurement of the Earth.

The only written work of Aristarchus that has survived is a little book called *On the Dimensions and Distances of the Sun and Moon.* In it he describes the way he went about trying to determine these dimensions and distances and the results he got.

The book begins with six "hypotheses":

1. The Moon receives its light from the Sun.

2. The Moon moves as though following the shape of a sphere and the Earth is at the central point of that sphere.

3. At the time of "half Moon," the great circle that divides the dark portion of the Moon from the bright portion is in the direction of our eye. In other words, we are viewing the shadow edge-on.

4. At the time of "half Moon," the angle at the Earth as shown in figure 1.3 is 87°.

5. The breadth of the Earth's shadow at the distance where the Moon passes through it during an eclipse of the Moon is the breadth of two Moons.

6. The portion of the sky that the Moon covers at any one time is equal to one-fifteenth of a sign of the zodiac.

Aristarchus's fourth and sixth assumptions are both far from accurate. The actual angle at the Earth in Aristarchus's triangle would be 89° 52′, not 87°, and 89° 52′ is very close to 90°. The angle at the Moon in Aristarchus's triangle *is* 90°. (See figure 1.3.) That makes lines B and C so close to parallel that, on a drawing, the triangle would close up and be no triangle at all. The portion of one sign of the zodiac that the Moon covers is not one-fifteenth, and it isn't clear why Aristarchus, who must have known this from obser-vation, chose that value.

By Aristarchus's calculation, the distance to the Sun turned out to be about nineteen times the distance to the Moon, and the Sun nineteen times as large as the Moon. Modern calculations put the

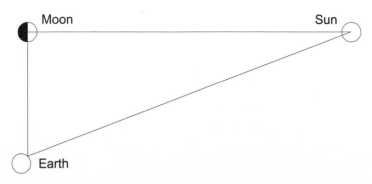

Figure 1.3 Aristarchus's measurement of the relative distances to the Moon and the Sun. When the Moon is a half Moon, the angle at the Moon (in this triangle) must be 90 degrees. So a measurement of the angle at the Earth determines how the length of the Earth-Moon line compares to the length of the Earth-Sun line; in other words, the ratio of the Moon's distance to the Sun's distance.

ratio between their distances at four hundred to one. The measurement Aristarchus was trying to make was extremely difficult with the instruments available to him. It is no simple undertaking to determine the precise centers of the Sun and the Moon or to know when the Moon is exactly a half Moon. Aristarchus chose the smallest angle that would accord with his observations, perhaps to keep the ratio *believable*. Throughout antiquity and the Middle Ages, estimates of the relative distances to the Sun and Moon would continue to be too small.

Aristarchus didn't stop with estimating the ratios but found ways of converting them into actual numerical distances to the Sun and Moon and diameters for both bodies. He could see that the *apparent* size (meaning the size a body *appears* to be when viewed from the Earth) of the Moon and that of the Sun are about the same, for during a solar eclipse, the Moon just about exactly covers the Sun. In more technical language: both bodies have about the same "angular size." Angular size tells how much of the total sky a body "covers" when viewed from Earth, and is measured in terms of "degrees of arc." The Sun and the Moon cover about the same amount of sky. They both have angular sizes of about one half of a degree of arc. A fuller explanation of those terms is in figure 4.4. For now, it is important only to know that the two bodies don't actually have to *be* the same size in order to have the same angular size, for how large they appear when viewed from the Earth (and how much of the total sky they cover) also has a great deal to do with how distant they are. See figure 1.4a.

Aristarchus correctly assumed that in spite of having the same angular size as the Moon, the Sun is actually much larger, and also much larger than the Earth. He knew that if the Sun was indeed much larger than the Earth, that made it safe to assume also that the shadow cast by the Earth has about the same angular size as the Sun and the Moon (one half of a degree of arc). Figure 1.4b shows what is meant by "the shadow cast by the Earth" and its angular size.

Aristarchus arrived at his fifth hypothesis (the breadth of the Earth's shadow at the distance where the Moon passes through it during an eclipse of the Moon is the breadth of two Moons) by observing a lunar eclipse of maximum duration, which means an eclipse in which the Moon passes through the exact center of the Earth's shadow. He measured the time that elapsed between the instant that the Moon first touched the edge of the Earth's shadow and the instant that it was totally hidden. He then found that that length of time was the same as the length of time during which the Moon was totally hidden. He reasoned that the breadth of the Earth's shadow where it was crossed by the Moon must therefore be approximately twice the diameter of the Moon itself (figure 1.4c). If, as he thought, the angle formed at the point of the Earth's shadow was the same as the angular size of the Moon, that gave him only one distance at which to put the Moon where it would cover half of the area of the shadow.

a.
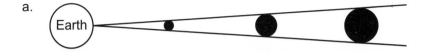

Surprisingly, all three of these bodies look the same size when viewed from Earth. We observe the "angular size" of a body like the Moon or Sun, not its true size. It could be small and close or large and far away and still have the same "angular size." Aristarchus saw that the Moon and the Sun have about the same angular size; that is, they *look* the same size when viewed from Earth, but he knew they are not the same true size.

b.   (The angles shown in this drawing are much larger than those that really exist.)

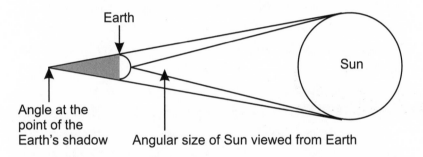

Earth

Sun

Angle at the point of the Earth's shadow   Angular size of Sun viewed from Earth

Aristarchus assumed that the Sun is much larger than the Earth. If that is true, then the angle at the point of the Earth's shadow is about equal to the angular size of the Sun as viewed from Earth.

c.

Observing an eclipse, Aristarchus concluded that the breadth of the Earth's shadow where the Moon crossed it was approximately twice the diameter of the Moon. He knew the angle formed at the point of the Earth's shadow and also the angular size of the Moon. There was only one distance to put the Moon where it would cover half the area of the shadow.

Figure 1.4 Aristarchus's calculation of the size and distance of the Moon. [Note: These drawings are not to scale.]

Aristarchus concluded that the Moon was one-fourth the size of the Earth, and that the distance to the Moon was about sixty times the radius of the Earth. Both of those values are close to the modern values. Using Eratosthenes's calculation of the Earth's radius, Aristarchus arrived at an actual distance to the Moon in stades. He had less success with the distance to the Sun. His earlier estimate—that the Sun's distance is about nineteen times the Moon's distance—was in error, and a second approach he tried, though it was ingenious and correct, required timing the phases of the Moon with a precision impossible in the ancient world.

It was another of Aristarchus's ideas that secured his place much more firmly in the annals of astronomy. Hearing of it, one has a chilling sensation of stumbling into a prophetic vision. For Aristarchus suggested, seventeen centuries before Copernicus, that the Earth is not the unmoving center of everything but instead moves round the Sun, and that the universe is many times larger than anyone in his time thought—perhaps infinitely large.

For centuries it had been widely assumed that the Earth was the center of the universe. The accepted picture of the cosmos was a series of concentric spheres—spheres imbedded one within the other—with the Earth resting motionless at the center of the system. (See figure 1.5.)

Plato and his younger contemporary, Euxodus of Cnidus, had introduced this model, and Aristotle's model of the universe was a further development of it, though he differed from Euxodus as to the number and nature of the spheres. However, it wouldn't be correct to think that everyone, without exception, since the dawn of human thought, had agreed that the Earth was the center and didn't move. Some Pythagorean thinkers had decided in the fifth century B.C., largely for symbolic and religious reasons, that the Earth was a planet and that the center of the universe must be an invisible fire. Heraclides of Pontus, a member of Plato's Academy under Plato,

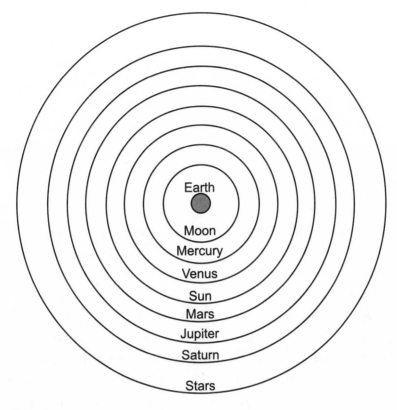

Figure 1.5

proposed that the daily rising and setting of all the celestial bodies could be nicely explained if the Earth rotates on its axis once every twenty-four hours.

But Aristarchus went farther. What we know about his theory of a Sun-centered cosmos comes secondhand; yet no one disputes his authorship of the idea because there is plenty of secondary evidence. According to Archimedes:

> *Aristarchus of Samos brought out a book of certain hypotheses, in which it follows from what is assumed that the universe is many times greater than that now so called. He hypothesizes that the fixed stars and the*

Sun remain unmoved; that the Earth is borne round the Sun on the
circumference of a circle . . . ; and that the sphere of the fixed stars,
situated about [that is, centered on] the same center as the Sun, is so
great that the circle in which he hypothesizes that the Earth revolves
bears such a proportion to the distance of the fixed stars as the center of
the sphere does to its surface.

Aristarchus had done no less than move the center of the cosmos to the Sun. In this astounding turnabout, the Earth moves
around the Sun and, rather than the sphere of the fixed stars making a revolution of the heavens once every twenty-four hours, it is
the Earth that turns, rotating on its axis—as Heraclides had suggested. The stars are extremely far away. The implication is, perhaps
infinitely far.

Did Aristarchus also speculate that the other planets move
around the Sun? There is no surviving evidence that he did or that
he understood the enormous significance of his model: that it provides, at a sweep, the basis for explaining the planets' positions and
movements far more simply than a model with the Earth as center.
In fact, there isn't even any evidence to indicate whether Aristarchus
really was personally disposed to thinking the Earth moved around
the Sun or whether he made the suggestion merely for the sake of
argument, as in "let's just suppose for the moment that this is how
things work." Why did this revolutionary suggestion come at this
time and place in history? The simple answer may be that this was
an intellectual environment that encouraged one to make suggestions and put forward hypotheses, even hypotheses based on assumptions that were known to be incorrect—in no way claiming
they were true—as the starting point for an interesting line of inquiry.

It's also important to ask why this idea died aborning. Seleucus
of Seleucia, a Chaldean or Babylonian astronomer (Seleucia was

on the Tigris River) in the second century B.C., took Aristarchus's suggestion seriously—not merely as a hypothesis. Seleucus believed Aristarchus was right. However, it appears no one else did, although antiquity was by no means a dark age when it came to astronomy. There seems to have been almost no public reaction at all. Aristarchus's suggestion must have been too far removed from common knowledge and common sense to draw much popular attention. The historian Plutarch reports one comment from the Stoic Cleanthes (the Stoics were reputedly weak in natural science, even "antiscientific") that Aristarchus of Samos ought to be indicted on a charge of impiety for putting the "Hearth of the Universe" in motion. There is no record of anyone trying to take Cleanthes's advice. Some philosophers scolded Aristarchus for trespassing on an area of knowledge that was *their* sole domain. There were also complaints accusing him of undermining the art of divination.

As for astronomers, what mattered most to them was that there was no observational evidence whatsoever to support the vast distances to the stars that Aristarchus's scheme required, while there *was* observational and physical evidence that made his Sun-centered arrangement seem highly unlikely:

1. If the Earth moves around the Sun, observers on the Earth should see some variation in the positions of the stars when viewing from different points along the Earth's orbit. No such variation had been observed (nor could it be with the technology available at the time). Aristarchus saw that this objection wouldn't be valid if the stars are far enough away. He suggested that they are very far away indeed, perhaps even at infinite distance. (The fact that the positions of the stars do change as the Earth orbits—that there is "stellar parallax motion"—wasn't confirmed by observation until the mid-nineteenth century.)

2. If the Earth rotates on its axis, in fact, if it moves at all, this should have some noticeable effect on the way objects move through the air. Scholars realized that if the Earth rotates on its axis once every twenty-four hours, the speed at which any point on its surface is moving is very great indeed. So how could clouds, or things thrown through the air, overcome this motion? How could anything ever move *east?* Solid bodies moving through the air should in some way show the influence of the Earth's rotation, even if the surrounding air rotates with the Earth on the Earth's axis.

3. It's plain to see that heavy objects travel toward the center of the Earth. If this law applies to heavy objects everywhere, the center of the Earth must be the center of gravity for all things in the universe that are heavy. Furthermore, once a heavy object reaches the place toward which its natural movement sends it, it comes to rest. Applying this idea to the Earth leads inevitably to the conclusion that the Earth must be at rest in the center of the universe and that it cannot be moved except by some force strong enough to overcome its natural tendency. This argument was based on Aristotle's concept of "natural" places and "natural" movements. It is easier to see its validity if you realize that Aristotle thought of everything beyond the Moon being made up of something called aether, which was neither "heavy nor light."

4. The Sun-centered model did nothing to solve a problem astronomers had been trying to solve: the inequality of the seasons measured by the solstices and the equinoxes.

If Aristarchus tried to answer the second, third, or fourth of these objections, those answers have not been recorded.

It would be inaccurate, and unfair to Aristarchus's contemporaries, to say that his Sun-centered model was suppressed because of their ignorance and closed-mindedness. The fact is, his model was an inspired guess that modern technology and theory show has far more validity than he or his contemporaries could possibly have known. But there actually was nothing coming from observation at the time to recommend it over the accepted view of the universe— the Earth-centered view that had been around for hundreds of years and that would be brought to a high degree of sophistication three centuries later by Claudius Ptolemy. Earth-centered astronomy solved the problems of astronomy, as they were perceived at the time, *better* than Aristarchus's Sun-centered model. Aristarchus's idea was a seed sown far too early, in a season in which it could not possibly germinate and take root.

Hipparchus of Nicaea

The greatest astronomer of the ancient world, indeed one of the most skilled and important of all time, was Hipparchus of Nicaea, who was born in the northern part of Turkey in the second century B.C. Thanks to Alexander's conquests, Hipparchus had at his disposal a priceless collection of Babylonian astronomical records, including eclipse records spanning many hundreds of years. He put this inheritance to splendid use, meticulously comparing the positions and patterns of stars and planets over the past centuries with those he observed himself. Like Aristarchus and Eratosthenes, Hipparchus tried to find a way to calculate the distances and dimensions of the Sun and Moon. Using a new line of reasoning, he focused on the fact that there was no discernible change in the Sun's position against the background of stars when an observer moved from one point to another on the surface of the Earth. Hipparchus assumed that solar parallax, as such a change in position is called, was just below the threshold of visibility. His results were not close to modern calculations.

However, Hipparchus's other efforts were far more successful. In comparing his own observations with those made about 160 years earlier, he discovered a change in the relative positions of the equinoxes and the fixed stars. That is, if you look at the stars on the evening of the spring equinox, and then again on the evening of a spring equinox some years later, the stars will not be in the same position. In fact, they won't be in the same position again for 26,000 years! This phenomenon is known as the "precession of the equinoxes." Though Hipparchus couldn't discover its cause, he gave an impressively accurate estimate of the rate of this change.

Of all Hipparchus's writings, only one youthful, minor work survives. Information about his accomplishments comes only from the references of others, mainly Ptolemy, but that information is sufficient evidence that Hipparchus was an extremely fine astronomer and that he vastly improved observational techniques, laying a foundation for all future astronomy. Where did this man stand in the competition between Aristarchus's model of the universe and the traditional one? Definitely pro-traditional. Hipparchus was among those who did not accept Aristarchus's Sun-centered cosmos, and he influenced others to reject it. Hipparchus felt obliged to abide by the evidence of observation—observational astronomy was, after all, one of his fortes—and, clearly, observation didn't support Aristarchus and couldn't confirm the enormous distances required by the Sun-centered model. Hipparchus's own work contributed significantly to Ptolemy's later Earth-centered model of the cosmos. Some scholars even insist that Ptolemy's astronomy was by and large a reediting of Hipparchus's, that Hipparchus was the genius and Ptolemy the textbook writer.

It seems fitting to close this discussion, and take leave of the ancient world, with a quote from the Roman Pliny the Elder:

Hipparchus did a bold thing, that would be rash even for a god, namely to number the stars for his successors and to check off the constellations by name. For this he invented instruments by which to indicate their several positions and magnitudes so that it could easily be discovered not only whether stars perish and are born, but also whether any of them change their positions or are moved and also whether they increase or decrease in magnitude. He left the heavens as a legacy to all humankind, if anyone be found who could claim that inheritance.

"If anyone be found . . ." ?

Heavenly Revolutions

{100–1600 A.D.}

> Now authorities agree that Earth holds firm her place at the center of the Universe, and they regard the contrary as unthinkable, nay as absurd. Yet if we examine more closely it will be seen that this question is not so settled, and needs wider consideration.
>
> **Nicolaus Copernicus, *De Revolutionibus Orbium Coelestium***

A few years ago, Harvard astrophysicist and science historian Owen Gingerich received a flyer in his mail offering a $1,000 prize for "scientific proof-positive that the Earth moves." A Mr. Elmendorf, who posed this challenge, wrote, "As an engineer, I am astounded that the question of the Earth's motion is apparently not 'all settled' after all these years. I mean, if we don't know *that*, what do we know?"

Indeed, whether the Earth moves can hardly be classed as one of the great unsolved mysteries of science. Schoolchildren learn that we live on a planet that revolves on its axis and orbits the Sun, that Nicolaus Copernicus introduced this controversial idea in the sixteenth century, and that some men were persecuted for believing it. But in the end . . . "all settled" . . . case closed. That was four hundred to five hundred years ago. What, you might ask Mr. Elmendorf, is the fuss about? And why has no one won the $1,000.

History and science turn out to be far more subtle and ambiguous than we're taught in the early grades.

Certainly no scientific knowledge has a better claim to being "Truth" than the knowledge that the Sun is the center of our planetary system and that the Earth, like the other planets, orbits it. Yet our own contemporary science backs away and tells us that when it comes to proving what moves and what doesn't, and whether or not there is an unmoving "center," no one can make an airtight case that any answer is right or wrong. Pick what you will, the Moon, Mars, the Sun, the Earth, your great aunt's dining table—the options are infinite—and it's possible to come up with a successful mathematical description of our planetary system with that as the unmoving center. In fact you are being parochial if you limit the exercise to our planetary system. It is possible to describe the entire universe using any chosen point as the unmoving center—the Earth will do very well—and no one can prove that choice is wrong. The issue here is one of relative motion only. You can measure the motion of an object only in relation to other objects in the universe. Scientists today prefer to picture everything in motion and nothing as being the center.

If you haven't given much thought to the implications of twentieth-century science, you may be as chagrined as Mr. Elmendorf to realize that because of the concept of relative motion, no one can prove that the Earth moves. Nor is relative motion Mr. Elmendorf's only problem. One tenet of science is that while an explanation can be extremely convincing and useful, none should ever be considered "final" or "proved," or "Truth" with a capital *T*. All scientific explanations are, in principle, open to revision and even complete rejection when better ones come along. Jules-Henri Poincaré, scientist and philosopher of science, was referring to this open-endedness of science when he wrote, "If a phenomenon admits of a complete mechanical explanation, it will admit of an infinity of other [mechanical explanations] which account equally well for all the peculiarities disclosed by experiment." Does this apply even to the motion of the solar system? Indeed, that was an example Poincaré used.

This deeper scientific insight notwithstanding, every generation including our own tends to believe devoutly in the finality of its own science, always for the same excellent reasons: What we experience presents us with puzzles. We put our trust in plausible solutions, in explanations that seem to make the best sense of things as we know them, and of things as we believe future generations will probably know them (to the best of our ability to predict). After all, that is the most anyone can ask of science in any era. But current scientific knowledge is never final, unassailable truth.

Carrying a modern worldview along on a visit to the past is notoriously inadvisable. Nevertheless, when that visit takes in medieval, sixteenth-century, and seventeenth-century astronomy, doing so selectively is not such a bad idea. Leave behind scientism, the popular impression that current science is, or claims to be, final Truth. Bring instead the less naive scientific worldview that recognizes the open-endedness of science. By all means, don't forget the concept of relative motion.

The Ptolemaic Carnival

Almost no information whatsoever exists about the life or personality of Claudius Ptolemaeus, known to us as Ptolemy, except that he worked at Alexandria during the second century A.D. and died in about 180. There is no record of where he was born, and his name doesn't mean he was a member of the ruling family. Like Eratosthenes, Ptolemy was interested in a wide range of subjects, including acoustics, music theory, optics, descriptive geography, and mapmaking. Some of his maps were still in use as late as the sixteenth century. More significantly, Ptolemy drew together, from previous ideas and knowledge and out of his own mathematical genius, an astronomy that would dominate Western thinking about the universe for 1,400 years.

Ptolemy inherited an intellectual tradition that placed an unmoving Earth at the center of the universe and that insisted that all heavenly movement occurs in perfect circles and spheres. Among his contemporaries who thought about such matters, most had concluded that the physical appearance of things must be taken into serious account when one tries to figure out the structure of the universe. To be believed, an explanation must "save the appearances." That may seem so obvious that it is hardly worth mentioning, but it isn't an assumption present in all cultures nor was it supported by all schools of thought in the Greek, Hellenistic, or medieval worlds.

The "appearance of things," for Ptolemy, included what Hipparchus and others had recorded in star catalogs. Ptolemy also brought to his task an in-depth knowledge of previous attempts to explain and predict planetary movement as it is seen from the Earth. The origins of the longing to understand that movement are lost in prehistory, but Plato, in the fourth century B.C., focused it in the question: "What are the uniform and ordered movements by the assumption of which the apparent movements of the planets can be accounted for?" The ancient attempts to answer him, beginning with his pupil, Eudoxus of Cnidus, were ingenious, and science historians disagree (as they do about Ptolemy's inheritance from Hipparchus) about how much Ptolemy's work was a synthesis of some of these earlier ideas and how much it was original with him. Either way, it's clear Ptolemy was a superb mathematician, and his achievement is almost unparalleled in the history of science.

When trying to understand Ptolemy, it helps to pay an imaginary visit to a fair or amusement park. It's night, and you are on a carousel designed for very young children. The horses move in a large circle and have no other motion. Put a light on the head of one horse and switch off all other lights, then situate yourself at the center of the carousel in such a way that you don't turn with the carousel. Now set it going. The light circles you steadily, never varying in speed or brightness, never changing direction.

If the Earth were the center and were not moving or rotating, and if all the planets were orbiting it in circular orbits, observers on Earth could expect to see each planet as you see the light on this carousel. That is, roughly, the way we see the Moon and the Sun, though their movements include irregularities that foil any attempt to describe them quite that simply. But we definitely do not see the planets moving in this manner, and neither did ancient astronomers and stargazers. Even as early as Plato and Eudoxus, those who studied the heavens knew that a model with simple circular orbits centered on the Earth couldn't adequately explain what was going on up there.

To illustrate one particularly mysterious problem, go back to the imagined carousel. Suppose you see the light move ahead for a while, pause, back up, then move forward again. The pattern continues to repeat itself. How to account for this movement? Someone might suggest that the light isn't attached to a horse's head at all. Instead it's on the cap of the ticket taker who is moving around among the horses. But the movement looks too regular for that. Not quite random enough. Try again. Maybe the light is at the end of a rope, and someone riding one of the horses is swinging the light around his or her head as one would a stone in a long slingshot

prior to launching it. Putting that movement together with the overall circular movement of the carousel might explain the apparent
backing up, as the light circles toward the back of the rider's head.
Or, if you want to stay with the notion that the light is on a horse's
head, perhaps the horse isn't fixed directly to the floor of the carousel but instead is part of a minicarousel attached near the edge of the
large carousel. In other words, in addition to being carried around in
the big circle of the carousel, the horse is also moving around in a
smaller circle, chasing its own tail. This last scheme would be something like figure 2.1. You could still accurately say that you are at the
center of the carousel and everything on it is moving around you.
Also, all movement is in perfect circles, no matter how complicated
and uncircular it may appear to your eyes.

Figure 2.2 is an idealized picture of the pattern such a light
might trace in time-lapse photography taken from a helicopter hovering over the carousel. However, from your position at the center
(E), you wouldn't see the loops. The light would appear to go forward, then pause, then back up, then pause, then repeat the pattern.
With the carousel fully illuminated, it's easy to account for that
movement in terms of perfect circles, but it would require considerable mathematical insight to do so if you could see nothing but the
moving light.

With the naked eye and a view of only a portion of the sky at
any one time, those who studied the movements of the stars, Sun,
Moon, and planets before the invention of the telescope found
themselves facing much the same dilemma you face looking at a
mostly darkened amusement park with no prior knowledge of the
rides that are there. Tiny lights that move smoothly for a time, then
pause and back up, or seem to grow brighter and dimmer, to speed
up and slow down. That's what Aristotle, Plato, Aristarchus, and
Ptolemy saw in the heavens. That was also what Copernicus saw,
for although he lived many centuries later than they, he too pre-

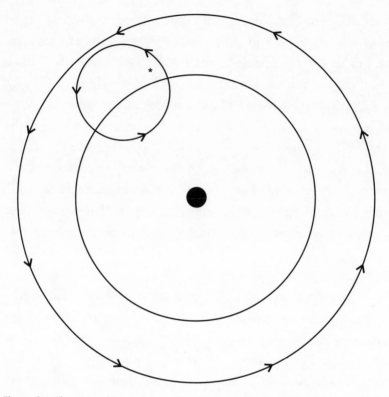

Figure 2.1 The carousel is rotating, and, on its periphery, the smaller disk is rotating on its own axis so that the horse with the light on its head (asterisk) is chasing its tail. From the center of the carousel it seems to us that the light moves forward, stops, reverses its motion a little while, stops again and moves forward once more.

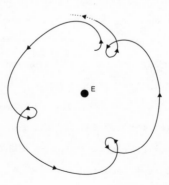

Figure 2.2

dated the telescope. Some of these men had charts and catalogs of what their forebears had observed, but even assuming those notations were made with consummate skill and care, given the lack of standardization and the discrepancy of calendars between one society and another, the interpretation of those records was not a straightforward matter. What a daunting task—one might think an utterly impossible task—to make sense of it, to design, as it were, carnival rides that could underlie and explain all that motion.

The marvel is that Ptolemy succeeded. The arrangement with small carousels riding on larger carousels is one of the models that he (long before such rides existed) took from his predecessors and incorporated into his astronomy. When we refer to planets rather than lights on carousel horses, the backing up movement is "retrogression." The smaller circles around which the planets move (chasing their tails) are "epicycles." The large circle around which the centers of the epicycles move is the "deferent." Figure 2.3 makes these definitions clearer.

By adjusting the size, direction, and speed of the epicycles, it's possible to account for many observed irregularities in the way the planets move as viewed from the Earth in an Earth-centered universe. It's also possible to explain irregularities in the movements of the Sun and Moon and variations in a planet's brightness. The epicycles place the planet sometimes closer to Earth, sometimes farther away.

A second device that Ptolemy inherited from earlier astronomy had a planet or the Sun orbiting in a circle with the orbit centered not precisely on the Earth but on a point a small distance away from the Earth, as in figure 2.4. The technical name for the displaced circle is the "eccentric." The planet, in this model also, will be closer to Earth in one part of its orbit than in another, which makes sense of variations in its brightness when viewed from the Earth, and also of apparent changes in its speed.

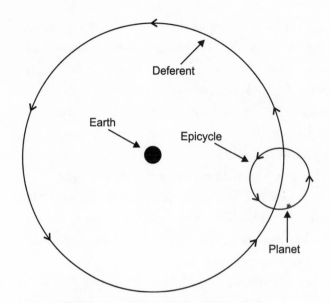

Figure 2.3 The planet moves in a small circle (epicycle) while that small circle moves along a large circle (deferent). The deferent is centered on the Earth.

Ptolemy originated a device called the "equant"—an imaginary point in the opposite direction (as viewed from the Earth) from the center of the "eccentric." This scheme kept the motion of a planet uniform, with respect to the "equant," while from the Earth its speed appeared to vary. Using equants, Ptolemy was able to predict the changing speed of a planet (as seen from the Earth) almost as successfully as Johannes Kepler would do in the seventeenth century. In fact, Ptolemy's solution came close to being the geometric equivalent of Kepler's.

Ptolemy combined these devices—epicycle/deferent, eccentric, equant—in an elaborate and highly successful model of heavenly movement. With astounding accuracy, his astronomy predicted and accounted for the observed movements of the five planets known in his time and of the Sun and the Moon, without removing the Earth from its position as unmoving center. Ptolemy centered the seven

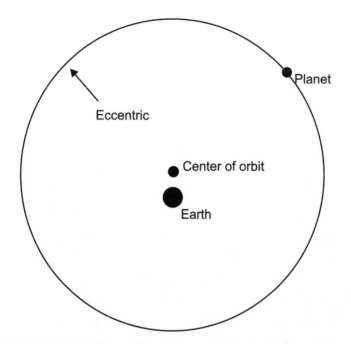

Figure 2.4 In this model, the circular orbit (the eccentric) is centered on a point a small distance away from the Earth.

orbits not on one point but on seven different points near the Earth. Adjusting epicycles, deferents, eccentrics, and equants, he succeeded in doing what his forebears had hoped for for so long—devising a system that could explain all the observed movement in the heavens in terms of spheres and circles.

When ancient and medieval astronomers, and also early Copernicans, spoke of "spheres," they weren't referring to the planets themselves. Since Aristotle first introduced the idea, long before Ptolemy, most astronomers had visualized the movement of each planet as taking place within the confines of its own invisible "crystalline sphere," though not everyone agreed on the nature and mechanics of these spheres.

In order to understand this concept, instead of imagining a flat drawing of circles within circles (such as figure 1.5), picture a three-dimensional glass object made up of transparent (invisible) spheres-within-spheres, like bubbles within bubbles. You can think of figure 1.5 as a cross section of such an object. In the Ptolemaic arrangement, the Earth, at the center, is surrounded by a sphere within which the Moon moves, and both of these are surrounded by a larger sphere representing the level at which Mercury moves, which in turn is surrounded by a still larger sphere representing the level at which Venus moves. Larger and larger nested spheres represent the levels at which the Sun and each of the other planets move. The outermost sphere represents the level of the stars.

In Ptolemaic astronomy, each planet's sphere is just roomy enough for the planet to cartwheel along on its epicycles. However, each sphere leaves room for no *more* than that. Ptolemy took from Aristotle the notion that the cosmos is a plenum. In other words, it is "full" in the sense that there can be no empty spaces between the spheres of the planets. Unlike nested Russian dolls, or bubbles within bubbles, the arrangement leaves no area that is not part of a sphere. The borders of the spheres "touch." For instance, the sphere of Jupiter, defined by Jupiter's nearest and farthest distances from the Earth, has an "inner border" that is right up against the "outer border" of the sphere of Mars—and so on and so forth with each planet's sphere and the spheres of its neighbors. Ptolemaic astronomy saw the spheres as pushing one another along. Movement originating in the sphere of the stars was thus transferred to the spheres of the planets, causing their movement.

In his most famous work, the *Almagest*, Ptolemy was chiefly concerned with predicting the positions of the heavenly bodies, not their distances from the Earth. However, working from Ptolemy's planetary scheme in that book, it's possible to calculate the ratio between the planets' greatest and least relative distances from the Earth, and Ptolemy was later to insist these could be transformed into *absolute* distances. He did the calculation first for the Moon and found that the least distance from the Earth to the Moon was thirty-three times the radius of the Earth (a radius known from Eratosthenes's measurement), and the greatest distance sixty-four times the Earth's radius. In a work that came after the *Almagest*, called *Planetary Hypotheses*, Ptolemy applied the same thinking to the planets and to the Sun, taking the scheme that had been largely of mathematical significance in the *Almagest* and treating it as "real" in terms of absolute distances. Ptolemy was unable to make it all work out without any glitches, but a strong argument in favor of his method nevertheless was that, with only small discrepancies, the results were largely in agreement with results of the study of eclipses.

As Ptolemy summed up his findings: The distance to the borders of the region "of air and fire" (the "sublunar," or lower-than-the-moon, region) is 33 times the radius of the Earth's surface ("the spherical surface of earth and water"). At that radius, the sphere of the Moon begins. Its farther border is 64 times that of the radius of the Earth (this was roughly where Aristarchus placed the Moon), and that is where the sphere of Mercury begins. And so on from there. The outer border of the sphere of Mercury is 166 times the Earth's radius. Venus, 1,079 times; the Sun, 1,260 (which agreed with Aristarchus's erroneous measurement to the Sun); Mars, 8,820; Jupiter, 14,189; Saturn, 19,865. That last radius is where the sphere of the fixed stars begins. When it came to the absolute distance to the fixed stars, Ptolemy wrote, "The boundary that separates the sphere of Saturn from the sphere of the fixed stars lies at a distance of five myriad myriad and 6,946 myriad stades and a third of a myriad stades." That comes out to 569,463,333 stades. Translated into modern miles, roughly 50 million miles. Today the measurement to the nearest star is over 25 trillion miles.

The term "Ptolemaic astronomy" is a bit of a misnomer. It

doesn't refer to a specific set of solutions coming from Ptolemy himself or any one of his successors. Instead, it means the combination and recombination of Ptolemy's devices as astronomers used them over the centuries. It is ironic to note that modern physicists and astronomers—who believe so firmly in a moving Earth—can recognize the power of these devices better than sixteenth-century Ptolemaic astronomers could. It takes understanding acquired through Copernican astronomy, as well as more highly developed mathematics than were available to ancient and medieval scholars, to appreciate how successful the older system potentially was in explaining the observed movement of the heavens.

Ptolemy's work wasn't lost with the decline of ancient civilizations. It had reached Baghdad in the eighth century A.D., and there it was translated into Arabic. The *Almagest* is an Arabic title meaning "the Greatest." From the eighth to the thirteenth and fourteenth centuries, the development of "Western" mathematics, astronomy, and astrology (astronomy and astrology were one and the same discipline then and continued to be even as late as the seventeenth century) took place not in Latin Europe but in the Middle East, North Africa, and Moorish Spain. Most scholars there who were responsible for this growth were Islamic, but not all, for these societies were cosmopolitan and tended to be tolerant of non-Islamic thinkers. Mathematicians and astronomers moved beyond the methods of the Hellenistic scholars, constructing observatories in Baghdad, Cairo, Damascus, and other leading cities. These observatories contained no telescopes, of course, but they did have intricate apparatuses—some inherited from the ancients and others that were newer inventions—to help plot the movements of the planets. Astronomical instruments came both in large bulky sizes, made of masonry, and in smaller portable versions.

Islamic astronomers seem never to have questioned the Earth-centered model. However, they did criticize Ptolemy on details, par-

A medieval woodcut of Islamic astronomers at work. *(Corbis-Bettmann)*

ticularly for the use of the equant, which, of all his devices, seemed nearest to violating the requirement of uniform circular motion. When Ptolemaic astronomy later became a well-established field of study in European centers of learning, European scholars did not forget their debt to Islamic scholarship. Copernicus mentioned some of his Islamic scientific forebears by name in his book *De Revolutionibus.*

There were attempts among Islamic astronomers to estimate the distances to the planets and the stars. One man who tried was Al Fargani, a ninth-century Arab astronomer who tried to gauge the sizes of the spheres and worked out relationships among these sizes. As Ptolemy had done, he allowed each sphere to be large enough to contain its planet's epicycles. Starting from a measurement of the Earth's radius of 3,250 Roman miles, he used these relationships to calculate the distances to all the known planets and to the sphere of the stars. His measurement of the distance to the stars had them more than 75 million miles from the Earth, half again as far as Ptolemy put them. The modern distance measurement to the nearest star is in the neighborhood of 1 million times greater than Al Fargani estimated to the sphere of the stars, but 75 million miles from the Earth was still a huge distance, and the sublunar (lower than the Moon) region in Al Fargani's universe was tiny indeed by comparison.

Ptolemaic astronomy, preserved and improved by Islamic scholars, filtered slowly into Latin Europe beginning as early as the eleventh century, but European scholars were much quicker to assimilate other ancient thinking and much more strongly influenced by it. The translation of Aristotle into Latin in the twelfth century had a profound impact. Scholars came to revere him not just as a philosopher but as "the Philosopher"—no other identification required—and the final authority on science and cosmology. With the passage of time, Aristotelian cosmology as interpreted by medieval scholars merged with medieval Christian thought and, somewhat later, with Ptolemaic astronomy—which agreed with Aristotle in many respects but not all. Scholars, who at that time were also clergy, after a period of disagreement and debate settled on ways to reconcile the Bible with Aristotle. In order to accomplish this, they gave Scripture a less literal, more metaphorical reading. They also eventually succeeded, to their satisfaction, in resolving contradictions among Ptolemy, Aristotle, and other ancient scholars to produce a coherent body of philosophical, religious, and scientific thought.

Aristotelian/Ptolemaic astronomy provided a visual, geometric structure for abstract medieval Judeo-Christian concepts, with all the rest of creation centering on that stationary Earth that is the home of humankind—a singularly unsavory place in Aristotle's philosophy, fallen from God's grace in Judeo-Christian teaching, and minuscule according to Al Fargani. By the thirteenth century, educated Europeans were taking it for granted that this worldview represented reality, and Dante, in the fourteenth century, both reflected and reinforced it in his *Divine Comedy*. He described a descent through the nine circles of hell toward the center of the Earth—the vilest point in the universe—and an ascent through the celestial spheres in which the planets move, to the throne of God.

By the time Copernicus was born in the fifteenth century, the simple Aristotelian picture of the cosmos and complex Ptolemaic astronomy (as received from the Arabs) had been mingling for at least three hundred years in the minds of European thinkers. Decked out in poetry and metaphor by Dante, this Aristotelian/ Ptolemaic/Judeo-Christian view of the universe now not only described and predicted heavenly movement with reasonable accuracy and served as a map of the physical universe, but had also come to symbolize the human condition and the geography of the spiritual universe: Among all creatures, only humans, fallen though they were, combined the material and the spiritual. Humanity was torn between the two, stuck on the debased squalid central Earth but always within view of and reaching out for the holy, pure, and changeless realms beyond.

However, the notion that the Aristotelian/Ptolemaic picture of the universe might be wrong was not completely absent from European thought. In the fourteenth century, it took the form of a challenge to Aristotle, not to Ptolemy, when Parisian Nicole Oresme wrote a commentary criticizing Aristotle's conclusion that the Earth doesn't move. Oresme didn't propose that the Earth moves in orbit. There is no suggestion that he thought of that. Actually, he didn't even conclude that it rotates, only that Aristotle had not proved it *doesn't*. A century later, Cardinal Nicholas of Cusa also suggested that the Earth was not lying motionless in the center of the universe. Nicholas of Cusa, however, failed to suggest another center. The next and far stronger challenge to Aristotle, Ptolemy, and Dante was to come from Copernicus.

Nicolaus Copernicus—
probably copied from his
self-portrait. *(Mary Lea
Shane Archives, Lick
Observatory, University
of California-Santa Cruz)*

Nicolaus Copernicus

Today the Polish city of Toruń on the banks of the Vistula River
has large modern buildings, but there also still exists a section of
Old Town whose narrow cobbled streets haven't changed much
since 1473, when Mikolaj Kopernik was born here. It was common
practice in scholarly circles to Latinize one's name. Mikolaj Kopernik became Nicolaus Copernicus.

Poland has had a tumultuous history and has often been divided
and dominated by foreigners, but the period during which Copernicus lived was a relatively peaceful, prosperous time. Copernicus's
father and grandfather evidently were merchants—well-off, if not
wealthy—and his mother came from a prominent Toruń family.
When Copernicus was ten, his father died and his mother's brother
took charge of Copernicus's upbringing and education. That uncle

rose to become bishop of Warmia, an influential position from which to advance the careers of his nephews.

When Copernicus was eighteen and his brother, Andrzej (Latinized to Andreas), was twenty, the two young men enrolled in the Jagiellonian University in Kraków, one of the most celebrated seats of learning in Europe and renowned for its astronomy. One of Copernicus's purchases while he was a student there was a set of the *Alphonsine Tables*. These tables, used for finding the positions of the Sun, Moon, and planets based on Ptolemy's theories and Islamic observations, had been computed in the thirteenth century and named in honor of King Alphonso of Castile. Copernicus's copy still exists in a collection in Sweden. Judging from the stains, he used it a great deal.

The story goes that when Copernicus was twenty-two, he and Andreas "walked across the Alps" to Italy to continue their studies. The influential uncle had taken his Doctorate of Laws at Bologna, Italy's oldest university and particularly famous for its Faculty of Law. He hoped to draw Copernicus's interest away from astronomy and into law. However, Copernicus didn't pass up the opportunity to get to know the leading scholars of astronomy and astrology at Bologna.

One eminent teacher there, Maria de Novara, not an astronomer but a mathematician, may have influenced him profoundly. Novara was a Neoplatonist, and Copernicus had already expressed some sympathy for Neoplatonic opinions. Neoplatonism was a framework of thought that stressed the need to discover simple mathematical and geometric reality underlying all the apparent complexity of nature. Neoplatonists, following in the tradition of Pythagoras, also saw the Sun as the source of all vital principles and energies in the universe. Novara himself insisted that nothing so complex and cumbersome as Ptolemaic theory could correctly represent the truth.

Copernicus was mathematician enough to find the complications of that astronomy unwieldy and annoying. But somewhere along the line, he went beyond mere annoyance and became thoroughly convinced that a simpler model of the universe was much more likely to be a correct model. It has been a puzzle in the history of astronomy why the challenge to Ptolemaic astronomy came when it did, at a time when there was no more compelling observational reason for it than there had ever been in the past. The answer to the questions "Why *now?*" and "Why from Copernicus?" may lie partly in Copernicus's association with Novara.

Due to their uncle the bishop's influence, Copernicus and his brother had received appointments as canons of Warmia, positions involving both Church and civil authority. Nevertheless, a leave of absence for educational purposes was cheerfully granted when Copernicus announced his intention to study medicine. Doctors were in short supply. So Copernicus set off again in 1501, this time for the University of Padua with its famed Faculty of Medicine. There he did indeed expend much effort studying medicine, but the degree he eventually received was a Doctorate of Canon Law from the University of Ferrara. The explanation for the change of venue may be that he didn't know anyone at Ferrara and hence could avoid the expense of the expected celebration party.

Copernicus returned to Poland in 1503 and—surprisingly, after such a cosmopolitan start—never left again. He began to build a distinguished reputation as a doctor, reportedly saving many lives later during a severe epidemic in 1519. His notes clearly reveal the primitive state of medical practice at the time, for Copernicus's knowledge was probably state-of-the-art.

From 1503 until 1512, Copernicus, now in his thirties, lived at Lidzbark Castle, the seat of his uncle, the bishop of Warmia. He served as his uncle's secretary and personal physician and became a force in the politics of Warmia, putting to use the legal education he had received at Bologna. Though his forte was never observational astronomy, Copernicus found time for observations and kept painstaking records of them. About 1507, while still a member of the court at Lidzbark, he wrote a book in which he claimed that the Ptolemaic model was wrong. The book was short, about twenty handwritten pages. In that form it circulated, at first anonymously and with an unrecorded title, among Copernicus's scientific acquaintances. The later title was *Nicolai Copernici de Hypothesibus Motuum Caelestium a se Constituis Commentariolus*, usually shortened to *Commentariolus* (Commentary).

Sun-centered astronomy as Copernicus first proposed it in *Commentariolus* didn't by any means solve all the problems still nagging Ptolemaic astronomy. Trying to eliminate the equant, which he found particularly offensive, and keeping the orbits circular, Copernicus could not avoid using epicycles to account for the movement of the planets. There is little mathematical reasoning in *Commentariolus*. It is hardly more than a sketch. Nevertheless, the leap that Copernicus made was extraordinary. He claimed that by putting the Sun in the center it would be possible to explain the heavens more simply and logically than Ptolemaic astronomy had done.

In *Commentariolus* Copernicus still visualized the universe in terms of spheres. The Sun rather than the Earth is at the center of his arrangement, and there is a sphere representing the level (from the Sun) at which the Earth moves, encasing the Sun and the spheres of Mercury and Venus. The small sphere in which the Moon moves is the only sphere that has the Earth as its center.

Copernicus proposed seven "assumptions":

1. All celestial spheres do not have only one common center.
2. The center of the Earth is not the center of the universe, but the center of the Earth *is* the center of gravity and of the Moon's sphere.

3. All spheres (except that of the Moon) revolve around the Sun, as though the center were the Sun, so the center of the universe is near the Sun.

4. The firmament of stars is extremely far away. The distance from the Earth to the Sun is insignificant when compared with the distance from the Earth to the firmament.

5. What appears to be the motion of the firmament of stars is not its motion but that of the Earth. The Earth with its adjacent elements, air and water, rotates daily around its poles, while the firmament remains motionless.

6. What appear to be motions of the Sun are not its motions, but the motion of the Earth and of the sphere with which it revolves around the Sun in the same manner the other planets revolve in their spheres.

7. What appear to be retrograde (backward) and forward movements of the planets are not their motions but the motion of the Earth. The Earth's motion alone is sufficient to explain many different phenomena in the heavens.

Copernicus went on to write: "The highest is the sphere of the fixed stars, containing and fixing location for everything. Below it is Saturn, followed by Jupiter, then Mars; below it the sphere in which we move, then Venus and finally Mercury. The lunar sphere revolves around the center of the Earth."

It would seem that Copernicus had delivered a bombshell, but there was no explosion. Few people outside Poland heard about his book, and his revolutionary proposals went largely undiscussed and unchallenged. The Catholic Church was aware of his work but didn't react either negatively or positively, adhering to a tolerant, hands-off policy it had been officially following for two centuries toward ideas that challenged traditional astronomy. If any dyed-in-the-wool Ptolemaic astronomers recognized Copernicus as a potential threat, they must have sagely decided that the best defense against the new model would be not to respond to it at all, thereby not calling attention to it. Let it die in oblivion.

In 1512, Copernicus's life changed with the death of his uncle. He left the castle at Lidzbark and went to live at Frombork, for he was still a canon of Warmia and Frombork was the seat of the Chapter, the assembly of the canons of the cathedral. There, in a tower adjoining the cathedral, he would do his most important work. He described his residence there as "the remotest corner of the Earth," which seemed not to bother him or unduly deflect him from his scholarship. Copernicus was a modest, quiet man with little objection to having himself, and the Earth, removed from the center of things.

During his years at Frombork, Copernicus continued to make astronomical observations, most of which were less accurate than those made by Hellenistic and Islamic astronomers. He also began or continued to write a second book. There were many interruptions. Copernicus became involved in negotiations with the Teutonic Knights, one of the forces that periodically threatened to tear Poland apart. When diplomacy failed and fighting broke out, Copernicus moved to Olsztyn, though he was too stiff-necked to retreat to Gdansk with the rest of the canons and remained at Olsztyn, as it came under siege, to organize resistance. When the siege was finally lifted and the fighting stopped, he headed relief operations. During the recovery after this conflict, Copernicus became involved with problems of economics, and he wrote a short, insightful tract proposing currency reforms. It's surprising to learn that had Copernicus not introduced Sun-centered astronomy, he might still be remembered as a minor historical figure in the field of economics.

Copernicus eventually got back to Frombork, where his duties as an administrator and doctor must still have consumed the greater part of his time. Nevertheless, he resumed work on his book. By now that may have been substantially completed, but he was a meticulous man, worrying over details. He was trying to confront and solve every problem presented in the astronomical data he had, and that was data coming from observations of insufficiently high quality even to show, in many cases, what the real problems were, much less to allow their solution. He was frustrated, for example, by the discovery that the Earth's whole orbit seemed to oscillate. He called these oscillations "trepidations" and tried to account for them. Later astronomers found that the trepidations were an illusion. Copernicus had been worrying himself over nothing but the result of bad data. In general, Copernicus was wrestling with a dilemma that has perennially beset astronomers and other scientists as well, not knowing which questions they can hope to answer and which not with the current technology and theoretical framework . . . sensing that not answering them all spells failure.

Copernicus resisted the pressure from his friends to publish his book while he was still unsatisfied with it. He was a well-known and respected astronomer and had no desire to appear an eccentric lunatic. He knew his vision of the universe disagreed with what nearly every intelligent, educated person had been thinking for millennia. He was personally convinced that his Sun-centered model could yield a far more harmonious and effective astronomy, but until he could demonstrate this clearly—with no loose ends—he did not want to publish. Though the issue was the large-scale arrangement of the universe—the big picture—he firmly believed that this was a battle to be fought in terms of technical and mathematical minutiae.

Copernicus had other worries dogging him. He was in his sixties now, no longer in the prime of health. Gnapheus, a minor playwright, produced a comedy, *The Wise Fool*, that mocked Copernicus. Some ill-considered, off-the-cuff, scoffing remarks about moving the

Earth, ostensibly coming from Martin Luther, were reported by one of Luther's dinner companions. Closer to home, Copernicus became embroiled in a demeaning squabble with the new bishop of Warmia. Reports differ as to whether Copernicus had opposed or supported this man's election, and whether it was a vindictive move when the bishop undertook to remove Copernicus's housekeeper Anna Szylling, a widow and distant relation on Copernicus's mother's side. She was a handsome, cultured woman whose presence in Copernicus's house the bishop (reportedly no paragon of morality himself) deemed suspect and unsuitable. Copernicus resisted for over a year but finally agreed to her departure.

Despite these griefs and distractions, in the late 1530s, 1,700 years after Aristarchus, Copernicus was at last drawing to a finish the book that ultimately would lead to the vindication of that ancient astronomer and be one of the most significant watersheds in human intellectual history: *De Revolutionibus Orbium Coelestium* (Concerning the Revolutions of the Heavenly Orbs). Copernicus wrote out the manuscript himself in longhand, as he had done with *Commentariolus*. This time there were more than two hundred pages. The book would elaborate on the sketch he had given in *Commentariolus*, his primary assertion again being that the Sun, not the Earth, must be considered the center of the system.

While Copernicus was still laboring on *De Revolutionibus*, the rumor got around that something radical was in the making at Frombork. The circulation of *Commentariolus* was small, but it was significant, and Copernicus's friends, particularly Bishop Tiedemann Giese, were spreading the word enthusiastically. Already in 1533, Pope Clement VII requested that his secretary explain these new Sun-centered theories to him. In 1536, Nicolaus Schönberg, Cardinal of Capua, wrote to Copernicus asking about his theories, and Copernicus sent him some explanations and tables. Cardinal Schönberg moved firmly into the Copernican camp. He urged Copernicus to allow his book to see the light of day and offered to pay for its publication and printing. Unfortunately, the Cardinal died before he could make good on his offer, but Copernicus mentioned this strong encouragement as well as that of Bishop Giese in the dedication of the book. However, it was not a Catholic churchman but a young mathematician named Rheticus, from Protestant Wittenberg, who at last persuaded Copernicus to publish his work.

Rheticus hadn't had an enviable adolescence. When he was a teenager, his father was beheaded as a sorcerer, and Rheticus, formerly Georg Joachim von Lauchen, changed his name. Rhaetia was the province where he was born. Now he was a junior professor at the University of Wittenberg, and he was deeply impressed with what he heard about Copernicus's ideas. In 1539 he traveled to Frombork to meet Copernicus in person. Rheticus evidently didn't lack for courage, because Wittenberg, Rheticus's university, was the center of Lutheranism while Warmia, where Copernicus was a canon of the cathedral, was Catholic and profoundly anti-Lutheran. But the two men, one sixty-six, the other twenty-two, seem to have hit it off splendidly, and Rheticus's visit stretched on for two years. He paved the way for *De Revolutionibus* by publishing a short volume of his own summarizing Copernican theory. His book was favorably received. Finally, Copernicus agreed to publish.

The tale, as it continues, is a convoluted one. The favored version is that Rheticus had to return to Wittenberg before Copernicus was prepared to part with the manuscript of his book. Rheticus took with him only some early mathematical chapters. A little later, with Copernicus's consent, Bishop Giese sent the completed manuscript on to Rheticus, who took it to a Nuremberg publisher, intending to keep close watch over its printing. However, he was then appointed to a new position with a higher salary in Leipzig and delegated what remained of the proofreading to Andreas Osiander, a Lutheran clergyman who was more nervous about possible religious reactions than Rheticus was. The Catholic Church had had nothing to say one way or another, except for the support of Cardinal Schönberg and Bishop Giese, but there had been those reports of adverse remarks from Luther. Osiander urged Copernicus to protect himself by writing a preface saying that his theory was intended to be taken hypothetically, not as a truth claim. Copernicus refused. He dedicated his book to Pope Paul III, a scholar interested in science. Osiander decided to go ahead and write, himself, the preface Copernicus wouldn't write. He left it unsigned, probably because he feared that his own antipapal reputation would cast suspicions on Copernicus.

On May 24, 1543, about a month after the printing of *De Revolutionibus* was completed, Copernicus died. The story is that he saw the printed book. He had had a stroke and was bedridden, perhaps unconscious—so there is some doubt whether that story is true, or, if it is, whether he was aware enough to realize what he was seeing. Did Copernicus find out that Osiander had written, anonymously, the preface containing the warning: "Beware if you expect truth from astronomy lest you leave this field a greater fool than when you entered"? If he did, there is no record of his reaction.

Moving the Earth

Even after Copernicus's prodigious effort and foot-dragging about publication, *De Revolutionibus* did not make the case for Sun-centered astronomy as effectively as he had hoped. There were many loose ends. Though he was able to eliminate the use of the equant, which pleased him, he still had to use epicycles and eccentrics to explain the movements of the planets. The result was hardly less complicated and cumbersome than Ptolemaic astronomy.

However, the new system clearly had some things going for it:

The rearrangement of the heavens allowed Copernicus to come at the problem of the mysterious backing up, or retrograde, movement of the planets in a new way. Retrograde movement occurs when a planet is in opposition, meaning that it is on the other side of the Earth from the Sun. (Only Mars, Jupiter, Saturn, and the other outer planets discovered after Copernicus's time can be in opposition. Venus and Mercury, whose orbits are closer to the Sun than Earth's, can never be in opposition.) Most of the time, the planets move from west to east against the background of stars. However, around opposition a planet appears for a while to move from east to west. Ptolemy had used epicycles to solve this problem. In the Copernican model, with all planets including the Earth orbiting the Sun, when one of the planets is in opposition, the Earth catches up and runs ahead of the other planet.

For an analogy, imagine two race cars, one on an inner track and the other on an outer track. You are riding in the one on the inner track. The stadium is completely dark except for a light on top of the car on the outer track and a few distant streetlights far beyond that. When your car catches up with that car and moves on ahead, the light will appear to you (against the background of distant streetlights) to backtrack. If the motion of your car is so constant and smooth that you believe you are standing still, you'll conclude that the other car has stopped for a moment, backed up, stopped again, and continued its forward motion. (See figure 2.5.)

De Revolutionibus also made sense of the fact that Mercury and Venus never stray far from the Sun. Ptolemaic astronomy had used deferents and epicycles to unravel this mystery. In Copernicus's model, with the orbits of Mercury and Venus lying within the Earth's orbit (because they are closer to the Sun than the Earth is), there is no mystery. Observers on Earth couldn't possibly see these planets anywhere else but near the Sun. This explanation of the orbits of Mercury and Venus was one success of Copernicus's model that many of his contemporaries could immediately appreciate.

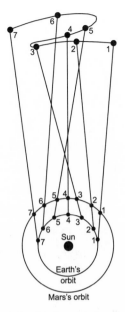

Figure 2.5 Copernicus's explanation for the retrogression of a planet. The Sun is in the center. The inner ring is the Earth's orbit. The outer ring is the orbit of the planet. Place yourself at #1 on the Earth's orbit, then at #2, and so forth, as the Earth moves in its orbit. The line drawn through the corresponding number on the planet's orbit shows where the planet is in your line of sight in each instance. The squiggle at the top of the drawing shows the pattern these changes of position (of both Earth and planet) will produce against the background of distant stars, and why the planet will seem to "back up."

Copernicus also addressed the ancient objections to the ideas that the Earth rotates on its axis and moves in orbit. Judging from modern knowledge about the availability of certain books during Copernicus's lifetime, most scholars conclude that when he wrote *Commentariolus* he probably didn't know about Aristarchus's ancient suggestion. However, it's clear from statements in *De Revolutionibus* that by the time he wrote that book he had heard about it. Copernicus explained that the Earth carries its atmosphere with it as it spins, and insisted, as Aristarchus had done, that the fact that we observe no stellar parallax indicates that the stars are extremely far away. *De Revolutionibus* required that the distance to the "Stellarum

Fixarum Sphaera Immobilus" (the immobile sphere of the fixed stars) be seventy-five times as great as Al Fargani's estimated 75 million miles—leaving a vast amount of empty space between the sphere of Saturn and the sphere of the stars. Copernicus, unlike Aristotle, Ptolemy, and Ptolemaic astronomers, thought that the stars were stationary.

In 1551, eight years after Copernicus's death and the publication of his book, the first handy-to-use tables based on Copernican theory appeared. (Those in De Revolutionibus itself were not at all handy to use.) They were a noticeable improvement on Ptolemaic tables—partly because there had been no new Ptolemaic tables for a very long time—but they were far from completely accurate, for astronomers, including Copernicus, were still so dependent on Ptolemaic observations.

Undermining the Ptolemaic Universe

No period in the evolution of thought about the universe and humankind's place in it has been more complicated or more ultimately decisive than the century and a half following the publication of De Revolutionibus in 1543. This was the so-called scientific revolution, slow getting started and slow-moving compared with most political revolutions, but in many ways more profound than any of those have been. How amazing that the shot eventually heard round the world came from a narrowly technical book that only highly trained astronomers and mathematicians could understand, whose author didn't make his case very effectively. There was almost no reverberation at all at first—nothing to indicate that over the course of a century and a half, in the minds of an increasing number of people, an Earth-centered universe would become a Sun-centered infinite universe; and science, for many, would become the chief arbiter of truth.

One reason for the initial dearth of reaction was that even the literate, educated public lacked the expert knowledge to understand *De Revolutionibus*. Among those who could understand it, few thought Copernicus had meant his proposal as a truth claim. Osiander's anonymous preface, which many assumed Copernicus had written, had something to do with this impression, but also, in the Ptolemaic tradition, new scientific/mathematical theories were normally intended as useful models for making predictions about planetary positions, not proposals for changing humankind's view of reality. The happy result for Copernican astronomy was that most scholars had found some of Copernicus's mathematical techniques too useful to discard by the time any actual opposition to his central thesis emerged.

De Revolutionibus's infiltration of the scholarly world was a slow process, and its actual subversion of Ptolemaic astronomy was even slower. Historian John Hedley Brooke points out that it is possible to identify only ten people in the years between 1543 and 1600 as "pro-Copernican" to the extent of stating that they actually believed the Earth moves and orbits the Sun. Short of that, some accepted Copernicus's proposal that the Earth rotates on its axis but not his proposal that it is in orbit. Others were willing to have the planets orbiting the Sun, while the Sun itself orbited the Earth, carrying the planets along. This scheme came from the great Danish astronomer Tycho Brahe and also from Nicolai Reymers Baer (better known as Ursus, Latin for "bear"), the official mathematician of Holy Roman Emperor Rudolf II. There were accusations and counteraccusations of plagiarism between the two men. In view of the concept of relative motion, this model was not ridiculous at all. In fact it is the geometric equivalent of the Copernican model—a problem that would continue to dog Copernican astronomers until well into the next century.

As word of the book and its contents spread, not all the debate about *De Revolutionibus* took place on strictly astronomical or mathematical grounds or among people who understood it. Outside astronomy circles, many people unaccustomed to drawing the line between hypothesis and truth claim *did* think of it as the latter, and the negative reactions were often vehement. Copernicus's Sun-centered astronomy challenged a cosmic structure that men and women had taken for granted since centuries before Christ, and, particularly, deep reverence for Aristotle still pervaded the educated world. Copernicus had tried to make the Earth one of the planets, but Aristotle had said that those were made of a fifth element that was not to be found in the corrupt realm within the Moon's orbit.

Tycho Brahe gave the world some fresh cause to doubt the Aristotelian picture of the universe by demonstrating that a nova in 1572, the great comet of 1577, and five more comets during the next twenty years were all farther away than the Moon. This was a shocker, because Aristotle had taught that birth, death, and change took place only in the sublunar part of the universe, and that beyond the Moon were only the eternal and unchanging celestial spheres. Tycho's discovery might seem a strong vote for Copernicus, but it fit equally well with Tycho's concept of all the planets orbiting the Sun while the Sun orbits the Earth.

Though the problem of having humankind dethroned from the center of the universe would frequently be raised as an objection to Copernicus's system, it was also possible to see this move as an innocuous or even a fortuitous one: Copernicus had argued that the fact that observers on Earth detect no stellar parallax as Earth travels in its orbit must mean that the stars are extremely far away. So in the Copernican system humans are, after all, still very close to the center. For all intents and purposes, with the distances being so great, they are still *at* the center. Some called attention to the fact that the center of everything in the Aristotelian scheme was not such prime real estate after all. Beyond the Moon's orbit was perfection; beneath it, corruption, at the center of which was the Earth. It was the realm of degradation and change. Christianity had come to associate it with fallen humanity. Satan lived in hell at the core of the Earth. Why cling so tenaciously to the notion that we live in the armpit of the universe? How welcome to find ourselves well out of that, moving around, breathing purer air! There was even, reportedly, criticism of the Copernican system for elevating man *above* his true station.

The question of possible extraterrestrial life arose and became a topic in religious discussions and writing. If Earth is a planet, might not other planets have inhabitants too? Had they fallen from God's grace as humans had? Did Christ die for them as well? Though this problem was a matter for conjecture and concern, most religious scholars didn't see it as a reason to reject the Copernican system, though some very vocal opponents mouthed off about it a great deal. Another issue had to do with the fact that before Copernicus heaven was considered to be beyond the outermost sphere. Hell was deep within the Earth. In the new system, was hell hurtling around in orbit and Dante spinning in his grave?

For the remainder of the sixteenth century and in the first decade of the seventeenth, the Catholic Church didn't oppose Copernicus's theories. *De Revolutionibus* was read, discussed, and even occasionally taught at Catholic universities, and Catholic scholars used computations based on Copernicus's work to produce the new Gregorian calendar in 1582. On the Protestant front, Luther didn't follow through with any more hostile remarks. Calvin observed that the Holy Spirit "had no intention to teach astronomy." The anti-Copernican statements often attributed to Calvin are fictitious—a late-nineteenth-century invention. Unlike modern Christian fundamentalists, neither Luther nor Calvin claimed that the Bible was the sole authority in matters other than faith and conduct. Of the ten aforementioned Copernicans between 1543 and 1600, seven were Protestant, three were Catholic. Meanwhile, as the poet John

Donne sagely observed, "most men lived and believed just as they had done before."

As time passed and the end of the sixteenth century approached, Copernicus's solutions and techniques and Copernican tables began to seem more and more convincing and indispensable to successive generations of mathematicians and astronomers, and each generation was less wedded than the last to Ptolemaic assumptions. Owen Gingerich, with whom this chapter began, has made it a sleuthing project to track down existing early editions of *De Revolutionibus*, to catalog them and to see what scholars of the sixteenth century penned in the margins in the tradition of "glossing" that still survived from the Middle Ages. He has discovered that marginal comments, clarifications, and criticisms were passed from teacher to pupil, sometimes for several generations, branching out family-tree fashion. Gingerich writes in his book *The Great Copernicus Chase*, "These second-hand annotations reveal that even if Copernicus's revolutionary new doctrine failed to find a place in the regular university curriculum, a network of astronomy professors scrutinized the text and their protégés carefully copied out their remarks, setting the notes onto the margins of fresh copies of the book with a precision impossible by aural transmission alone." Undoubtedly Copernicus's ideas were spreading and having an increasing impact. It appeared as though Copernican astronomy were headed for a peaceful victory, and Scripture could and would be reinterpreted once again to agree with a new vision of the universe.

Why was Copernicus's book succeeding—even before there was any new observational data to support it—in overturning Earth-centered astronomy, while Aristarchus's proposal had failed? Partly because Copernicus was not the only scholar laboring under the cumbersome, accumulated burden of Ptolemaic astronomy and wishing for improvement. Also, while Aristarchus had simply raised the suggestion that the Sun, not the Earth, lay at the center of the universe, Copernicus gave his readers a good deal of mathematics to consider, mathematics that they found interesting and useful even if they didn't follow Copernicus all the way.

But to understand more fully why an idea that was ignored in the ancient world should finally make its impact seventeen centuries later, it's necessary to look beyond astronomy and mathematics. The sixteenth and early seventeenth centuries in Europe were an era of intense intellectual, religious, political, and cultural ferment. There was a mounting spirit of upheaval and distrust of old assumptions, across the board. Anti-Aristotelianism and humanism were challenging and infiltrating scientific thought with Neoplatonic preferences for geometric and mathematical harmony and simplicity that Ptolemaic astronomy could not provide. Luther and Calvin and their followers, and Henry VIII as well, were calling into question the ultimate ancient authority of the Roman Catholic Church, offering, in its place, not one authority and doctrine but a rich and confusing choice. The Catholic Church was also of many minds on many fronts, so that it is misleading to speak of "the Church," as though it was a monolith, holding one opinion. The great age of exploration was under way. Columbus had sailed for the New World when Copernicus was in his late teens. Ptolemy's ancient maps were proving to be inaccurate. Was there perhaps reason to distrust his astronomy as well? Ancient copies of Ptolemy's work had turned up and made it impossible to sustain the hope that problems with his astronomy stemmed only from Arabic misinterpretation that could be corrected if only scholars knew what Ptolemy had actually said himself. This was a world in many ways ready to entertain the exhilarating thought that with Copernicus humankind had finally not only caught up with, but surpassed, the thinking of the ancients and was ready to move onward, unintimidated by the past.

Nicolaus Copernicus didn't single-handedly overturn the Earth-centered view that had prevailed since ancient times. But he did crack the door that would lead our ancestors toward the modern understanding of the universe. This time, the door wouldn't be closed again. As 20th century astronomer Fred Hoyle puts it in his book *Nicolaus Copernicus: An Essay on His Life and Work,* "It is because Copernicus focused the attention of the world at precisely the right spot, the place where Nature simply had to give up her secrets, that today we judge his work to have been so important." Hoyle uses a mountaineering term: Copernicus discovered the *point of attack.*

Others were gearing up to make the climb.

3

Dressing Up
the Naked Eye

{1564–1642}

> ❝I do not feel obliged to believe that the same God who has
> endowed us with sense, reason, and intellect has intended
> us to forgo their use.❞
>
> **Galileo Galilei**

It would be difficult to imagine two educated men of the same historical period more different from each other in background and personality than Johannes Kepler and Galileo Galilei. Kepler was a quiet, introspective man from an obscure village on the outskirts of the Black Forest near Stuttgart. Galileo was a colorful, feisty, larger-than-life character who grew up in Pisa and Florence when Florence was a world center of wealth, political power, and artistic and intellectual ferment.

Kepler's ancestry was respectable enough—his grandfather had been burgomaster. But Kepler's father was an evil-tempered ne'er-do-well who abandoned his family, and Kepler's mother was a malicious village troublemaker who dabbled in the occult and barely

escaped being burned as a witch. Galileo's father was a well-educated trader and an accomplished musician and music theorist.

Kepler didn't win friends easily. An unassuming, private man,
he struggled to make a living teaching, which he didn't do particularly well, and casting horoscopes, which he reputedly did very well
indeed, while pursuing his real passions: mathematics, astronomy,
and philosophy. Galileo won both friends and enemies readily,
relished the spotlight, and lived a public, even celebrity life. He
thought highly of himself, and he had a talent for conveying his
scientific ideas to nonexperts, including some in high places whose
favor he curried.

Kepler was an exuberantly devout Protestant; Galileo, a staunch
Catholic.

But far more significant than any of these differences was the
contrast between the ways these two men approached their science.
Kepler's mind moved by leaps of fancy and intuition inspired by his
Neoplatonic Christian faith that the universe must have a beautiful
hidden harmony to it—that things as far apart as music and geometry and cosmology must have connections and explain one another.
His most celebrated discoveries seem like small islands of dazzling
insight in a sea of wild, woolly thinking. But though much of his
work appears crazy to modern eyes, his greatest contributions were
at first equally much flights of imagination and were equally motivated by his longing to uncover symmetry and relationships. It took
a mind that could think as "far out" as Kepler, follow as many false
leads as he, and then proceed to pin ideas down with conscientious
mathematical rigor, to discover connections that really do exist.

Galileo, on the other hand, started from what he observed, in
the firm conviction that the only way to learn the truth about nature
was to examine it directly and put it systematically to the test.
Though eager and able to speculate—sometimes too optimisti-
cally—about the implications of his findings, he was reluctant to

espouse publicly or even among friends ideas for which he didn't personally see clear support from his own experiments or observations.

For all their genius, both of these men were also remarkably favored by happenstance. Each had fall into his hands something without which his most important discoveries would never have been made. For Kepler, that was the naked-eye astronomical observations of the great Danish astronomer Tycho Brahe. For Galileo, it was the telescope.

Together, Kepler and Galileo were responsible for the triumph of Copernican astronomy, yet they never met in person.

"Skybound Was the Mind"

Johannes Kepler was born in 1571, twenty-eight years after the death of Copernicus and the publication of De Revolutionibus. His boyhood in Weil der Stadt, with his sinister mother and, occasionally, his irresponsible father—in a household full of other unsavory, unhappy relatives—could have come straight from a Charles Dickens novel. The young Kepler was seldom in good health, and a childhood illness left him with weak eyesight. In school, he wasn't good at making friends. Nevertheless, it became clear early on that he had extraordinary intellectual gifts and a deeply religious nature.

Kepler hoped to become a Lutheran minister, and he went on from the local school to a theological academy for the children of "poor and pious people" (Kepler's family met the first requirement, if not the second) and then to the university at Tübingen, where scholarships paid his way. The Senate of the University observed that Kepler had "such a superior and magnificent mind that something special may be expected of him." Tübingen still taught Ptolemaic astronomy, but it was during this period that Kepler became

Johannes Kepler. *(Mary Lea Shane Archives, Lick Observatory, University of California–Santa Cruz)*

a Copernican, perhaps due to the influence of the private views of the astronomer Michael Mästlin, who was one of his teachers. An excellent example of what astronomy was like before telescopes was Mästlin's study of the nova of 1572, using no instrument at all except for a piece of thread. His results were more accurate than anyone else's, including Tycho Brahe's.

In 1594, when Kepler was twenty-three and in his final year as a theology student, a teaching job in mathematics and astronomy opened up at a seminary in Graz, Austria. The University of Tübingen nominated Kepler. Though surprised and somewhat distressed at this sudden change in his career trajectory, Kepler accepted the position.

Kepler wasn't a good teacher, but his new job evidently did leave him time to pursue his science and philosophy, for only two years after he began teaching he published the first book since *De Revolutionibus* to defend Copernican theory, and he made a far stronger

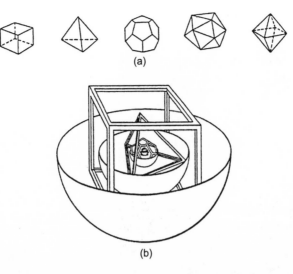

(a)

(b)

Figure 3.1 (a) The five regular or "cosmic" solids. In every case, all of the sides are identical and only equilateral figures are used for them. These are the only possible regular or "cosmic" solids. (b) Kepler's arrangement of them in relation to the spheres of the planets. Saturn's sphere is outside the cube. Jupiter's sphere is between the cube and the tetrahedron. Mars's sphere is between the tetrahedron and the dodecahedron. Earth's sphere is between the dodecahedron and the icosahedron. Venus's sphere is between the icosahedron and the octahedron. Mercury's sphere is within the octahedron.

case for it than Copernicus had been able to do. The twenty-four-word title of Kepler's book is usually shortened to *Mysterium Cosmographicum* (The Cosmographic Secret). *Mysterium Cosmographicum* elicits raised eyebrows from modern readers, in spite of the book's support of Copernicus and the ingenious nature of its proposal, for Kepler attempted to explain both the number of planets and the sizes of their orbits in terms of a relationship between the planetary spheres and the five regular solids of geometry: cube, tetrahedron, dodecahedron, icosahedron, and octahedron. (See figure 3.1.)

The publication of *Mysterium Cosmographicum* led to the first known contact between Kepler and Galileo. At Padua, where he was teaching at the time, Galileo received a copy of Kepler's book.

He wrote to Kepler saying he was looking forward to reading it and also that he had long believed in the Copernican theory himself but had not said so openly, to avoid ridicule. In a return letter, Kepler urged Galileo to make his opinion public. Galileo didn't take that advice.

Kepler continued eking out a living in Graz as an ineffective teacher, went on with his astronomical work, wrote some books about astrology, and married—a marriage that lasted, though there is some indication it was not an entirely happy one. Then in 1598, the Archduke Ferdinand began to make life miserable for Lutheran leaders and teachers. Things went from bad to worse, and eventually Kepler was given a day's notice either to leave Graz altogether or be sentenced to death. Kepler departed, temporarily, he hoped, for such persecution tended to wax and wane. But it soon became evident that the archduke's attitude was intransigent. Kepler would no longer be able to live and work in Graz.

However, Kepler was not completely out in the cold, for the opportunity had arisen to join Tycho Brahe at Prague. Kepler had sent the older man a copy of *Mysterium*, and Tycho had recognized a serious talent. Kepler was understandably apprehensive about how well the two of them would get along, for Tycho had a reputation as a proud, imperious eccentric. His nose had been partly sliced off in a duel, and he had restored the missing bit himself with gold, silver, and wax. But Tycho was also unarguably the greatest astronomer of his generation . . . and Kepler needed a job. So Kepler, now thirty years old, moved to Prague in 1601. Whatever the difficulties working with Tycho, he didn't have to cope with them long, for within two years Tycho died. Kepler succeeded him as Imperial Mathematician.

This new position was a considerable improvement over teaching at Graz, but there was a negative side. Although he had Tycho's impressive title, Kepler's salary was much lower than Tycho's had been and often it wasn't paid. Kepler was obliged to waste a great deal of time trying to collect what was due him. The job as Imperial Mathematician did, however, have other compensations, for, over the objections of Tycho's relatives and fortunately for the future of astronomy, it was Kepler who fell heir to Tycho's magnificent set of astronomical observations, the best the world had ever known. Tycho had found that circular orbits were difficult to reconcile with the actual paths of the planets, and he had undertaken an exhaustive series of observations that he hoped would throw more light on the problem. No man was better suited than Kepler to put this precious inheritance to optimum use.

Kepler brought to this work both a firm belief in Copernican astronomy and an unwillingness to accept that the skilled and meticulous Tycho's data could be faulty. The two must somehow fit, even though Tycho himself had rejected Copernican astronomy in favor of the scheme that had the Sun orbiting the Earth and all the other planets orbiting the Sun. Kepler also brought to his task a concept of his own that the Sun moves the planets by a "whirling force" centered in itself. Orbits centered elsewhere than on the Sun would not do.

Tycho Brahe had paid particular attention to the planet Mars, and it was in trying to make sense of these observations that Kepler finally found himself obliged to abandon circular orbits, removing the stumbling block that had hobbled Copernican astronomy since its inception. He realized that by using elliptical orbits he could explain Tycho's observations. Kepler signaled his overwhelming joy and astonishment at his insight by falling to his knees and exclaiming, "My God, I am thinking Thy thoughts after Thee." Those who describe Kepler as dry and passionless and his life as uniformly drab and sad simply fail to appreciate the sorts of things that moved him.

Kepler's discovery and Tycho's earlier observations that made it possible are both awe-inspiring achievements. Tycho's observations were made without benefit of a telescope, and Kepler discovered elliptical orbits in spite of the fact that the ellipse in which the Earth orbits is so nearly circular that any attempt to make it obviously an ellipse in a scale drawing ends up being a distortion. It is only slightly more obvious that the orbit of Mars is an ellipse. Yet with this seemingly trivial geometric alteration, the Copernican system fell into place. Kepler, aware of the power of his discovery and also of the reaction it would inevitably receive, commented wryly that with this introduction of elliptical orbits, he had "laid an enormous egg." He had indeed, and even many who admired his work found this egg difficult to digest. Galileo, for one, never accepted it.

The term *ellipse* isn't synonymous with *egg shape* or *oval*. To be more precise about it, think of an ellipse in one of two ways: One is as a slice out of a cone. (See figure 3.2.) Not just any oval or egg shape can be produced by slicing a cone, and so not all ovals and egg shapes are ellipses. But all horizontal and diagonal slices through a cone that don't go through the base of the cone produce ellipses. A circle, which is a slice directly—horizontally—across, is considered an ellipse.

A second way of thinking about an ellipse gets the same results. Imagine a piece of thick cardboard lying on the table before you. Stick two thumbtacks into it, any distance apart. Then take a piece of string that is longer than the distance between the two tacks and attach its ends to the tacks—something like the string shown in figure 3.3. Next take a pencil and pull the string taut with its point (figure 3.4), still keeping the string flat on the surface of the cardboard. If you move the pencil, allowing it to slide along the string while the string remains taut, the point of the pencil will draw an ellipse. Changing the length of the string or moving the thumbtacks closer together or farther apart changes the shape of the ellipse, just as changing the tilt of the slices in the cone does. The closer together the two thumbtacks are, the nearer the ellipse comes to being a circle. If they are in precisely the same location, it *is* a circle.

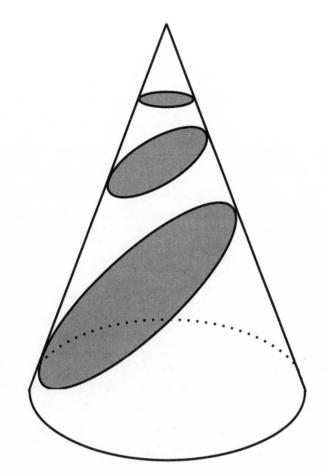

Figure 3.2

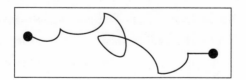

Figure 3.3

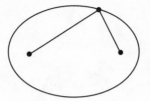

Figure 3.4

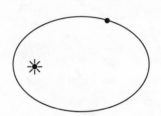

Figure 3.5

Each thumbtack represents a *focus* of the ellipse, so an ellipse has two foci. If a planet's orbit is near to circular, that means that the ellipse's foci must be very close together. By contrast, comets, which also orbit the Sun in elliptical orbits, have foci that are very far apart, producing an extremely elongated ellipse.

The first of Kepler's two Laws of Planetary Motion that appeared in his book titled *Astronomia Nova* (New Astronomy) states that a planet moves in an elliptical orbit, and the Sun is located at one of the two foci of that ellipse. (See figure 3.5.)

Kepler's second law has to do with the variation in the speed of a planet as it travels in its orbit. The law states that an imaginary straight line (called the radius vector) joining the center of the planet to the center of the Sun "sweeps out" equal areas in equal intervals of time.

Figure 3.6 shows an elliptical orbit, the Sun (one of its foci), and a planet traveling along in the orbit. Because a still picture can't

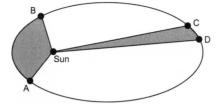

Figure 3.6

adequately demonstrate Kepler's second law, imagine that you are seeing a moving computer graphic. It begins with the planet at *A* and a straight line drawn from *A* to the center of the Sun. A stopwatch times the planet as it orbits to *B* (which for purposes of this demonstration could be located anywhere on the part of the orbit nearer the Sun). The imaginary line moves with it like the hand of a clock, shading the area it sweeps across. At *B*, the shading and the timing stop, and what you see on the screen is a shaded area that has been "swept out" as the planet traveled from *A* to *B*. The graphic at this point looks like the left side of figure 3.6. In the corner of the computer screen, a number appears to show how long it took the planet to travel that distance and sweep out that area.

The graphic shows the planet continuing in its orbit. When it reaches *C*, somewhere on the side of the ellipse farther from the Sun, another line appears, drawn from the planet to the center of the Sun. The shading and stopwatch begin again. The planet travels along and so does the sweeping line. When the planet has traveled for *the same amount of time it took to go from* A *to* B (the interval of time previously recorded in the corner of the screen), the shading stops. That point is labeled *D*. The computer graphic now looks like figure 3.6. Notice that on the left side of the picture, from *A* to *B*, the planet traveled a much greater distance than it did from *C* to *D* over on the right side, yet the interval of time, on the stopwatch, was the same. Clearly the planet didn't maintain a constant speed. Just as an automobile has to travel much faster to cover sixty miles in an hour than it would to cover thirty miles in an hour, the planet has to travel much faster from A to B, to cover all that distance, than it did to cover the smaller distance from C to D in the same interval of time.

Kepler discovered that the two shaded "swept out" portions are equal in *area*. Kepler's law thus predicts that a planet will always travel at a faster rate the nearer its path comes to the Sun. Kepler thought the variation in speed occurred because the distance from the Sun affected how much or how little the planet felt the Sun's "whirling force."

Kepler's second law, then, accounts for the way a planet varies in speed, and makes it possible to predict those variations. While you can't make sense of the system by saying "this planet always travels thus-and-so-many-miles per hour," you *can* make sense of it by saying "this planet's imaginary line to the Sun always sweeps out an area of thus-and-so-many-square-miles per hour." This was one of Kepler's far-out ideas, a connection he found between astronomy and geometry, arising from his belief in the intrinsic harmony of a universe created by God. Nearly every high school student today knows he got this one right.

Kepler's *Astronomia Nova* contained more than these two laws. He was well on track toward the modern way of thinking about gravity, previewing Newton's later discoveries about the tides and understanding this phenomenon a great deal better than Galileo did. Kepler had discerned that the more massive a body, the stronger its attractive effect. He also wrote that it couldn't be the Sun's light that provides the "whirling force," for the Earth doesn't screech to a halt during a solar eclipse. Kepler finished *Astronomia Nova* in 1606 but didn't publish it until 1609. That was the same year Galileo first looked through a telescope.

After the death of Kepler's wife and son, and with difficulties for Protestants mounting in Prague with the Counter Reformation as they had earlier in Graz, Kepler moved to Linz in 1611. He lived there for fourteen years, marrying a second time. It was in Linz that he published, in 1619, a book called *Harmonice Mundi* (Harmony of the World). According to this book, each planet has its own melodic line associated with its changing speed as it moves nearer to or farther from the Sun in its orbit. Those planets that have less variation in speed have more monotonous melodies. Kepler speculated that by working backward in time from current planetary positions and calculating the moment when the orbits of the planets would have produced the most perfect musical harmony, it might be possible to discover the moment of Creation.

More significant for future astronomy, *Harmonice Mundi* also contained Kepler's third law of planetary motion, which establishes a relationship between the lengths of time the planets take to complete their orbits and their distances from the Sun. Kepler summed up the relationship in the equation $(T_A/T_B)^2 = (R_A/R_B)^3$. T_A is the time it takes planet A to complete its orbit (its orbital period). T_B is the time it takes planet B to complete its orbit. R_A is the average distance from planet A to the Sun. R_B is the average distance from planet B to the Sun. The ratio of the squares of the orbital periods of the two planets is equal to the ratio of the cubes of their average distances from the Sun. The orbital periods could be timed and were known, but even so, Kepler's law didn't provide the means to calculate any *actual measurement* for the distance from any planet to the Sun. The third law was an equation in search of just one absolute, known distance that could be plugged into it to dictate the scale of the solar system.

Kepler's discovery of his first and second laws had given him a completely new foundation for putting together more accurate tables for computing the planets' positions at any time in the past or future, but the completed tables were slow in coming. Kepler's preface to the new tables explains that this was partly because of difficulties finding financial backing (he finally paid for the publication out of his own pocket) and also because of "the novelty of my discoveries and the unexpected transfer of the whole of astronomy from fictitious circles to natural causes." He went on to point out that no one had ever attempted anything of this kind before. It required intense study of Tycho's astronomical observations. Kepler tried orbit after orbit, leaving one after another behind when, after prodigious computation, they didn't match his data. He was not particularly happy doing this work. It must have been pedantic for so imaginative and creative a mind. It wasn't until 1627 that he finally published his *Rudolphine Tables*, and they were a triumph for Copernican astronomy.

Kepler spent his last years at Sagan, in Silesia. He died at Regensburg on yet another trip to try to collect back salary, in November 1630, just a year short of seeing that his prediction, in accordance with the *Rudolphine Tables*, that Mercury would cross the disk of the Sun on November 7, 1631, was correct. Kepler had composed his own epitaph:

> I measured the skies, now the shadows I measure
> Skybound was the mind, Earthbound the body rests.

The "Starry Messenger"

Compared with Kepler's bleak childhood, Galileo's apparently was a pleasant one in a family that valued intellectual pursuits. He was born seven years before Kepler, in 1564, in the north-Italian city of Pisa and attended school both there and in Florence. As a schoolboy he thought he would become a monk. However, at age seventeen, honoring his father's wishes, he entered the University of Pisa as a medical student. Two years later, when it was clear that mathematics and mechanics, not medicine, were Galileo's forte, his father allowed him to change his course of study.

One of Galileo's important discoveries took place during his student years at Pisa when he became fascinated by a lamp swinging on a long cord in the cathedral. Evidently he was not always completely attentive at church services. He noticed that regardless of whether the length (in space) of each swing of the lamp was long or short, the time it took to complete the swing seemed to remain the same. Curious, he experimented on his own and found that the length of time it takes a pendulum to complete one swing depends not on how big the swing is but only on the length of the cord by which it hangs.

In 1585, Galileo's formal education ended when his father ran short of money to pay university costs. Galileo came home to Florence, where his family was then living. Undeterred, he studied on his own and with scholars among his father's acquaintances. Soon he was making a local reputation with several inventions and discoveries, circulating his own short book on measuring the specific gravities of bodies, and brashly voicing suspicions about Aristotle's mental capacities. One bizarre undertaking during these years was a series of public lectures he delivered about the shape, size, and location of Dante's hell—a topic that might draw a considerable audience even today.

In 1589, when Galileo was twenty-five, four years after he had left the University of Pisa without a degree, he returned there as a lecturer. Pisa, like Tübingen, still taught Ptolemaic astronomy. Whatever Galileo may have been thinking privately, that was what he taught. It isn't clear exactly when he became a convinced Copernican.

A discovery Galileo made while he was a lecturer at Pisa caused his already low opinion of Aristotle to sink even farther. Though Aristotle had in truth been rather ambiguous on the subject, at least in those writings that survive, most scholars of Galileo's time, including Galileo, thought Aristotle had said that if two objects were dropped simultaneously from the same height, the heavier would strike the ground first. Galileo either dropped weights on numerous occasions from the Leaning Tower of Pisa, as his earliest biographer reported, or rolled them down a ramp, or perhaps he tried both. But he established to his satisfaction that the heavier and the lighter hit the ground at the same time. Galileo could not, of course, perform this experiment in the absence of air resistance. Science historian Thomas Kuhn has quipped that it probably wasn't Galileo who carried out the legendary public demonstration from the Leaning Tower but a defender of Aristotle, who thereby proved quite decisively for all present that Galileo was wrong and Aristotle was right. In the twentieth century, astronauts performed the experiment in the airless environment of the Moon. Galileo was right.

Never a tactful man, Galileo somewhat incautiously proceeded to debunk the still highly venerated Aristotle, saying he "wrote the opposite of truth" and was "ignorant." Galileo also showed a lack of judgment when he bluntly advised the grand duke of Tuscany, Ferdinand I (who had granted him his professorship), that a dredging machine designed by the grand duke's brother-in-law wasn't going to work. The machine was built nevertheless and *didn't* work, which made Galileo even less popular. When his father died in 1591, Galileo as eldest son was left responsible for the family. His salary at Pisa was far from sufficient and, thanks to the dredging machine incident, not likely to rise. He looked for a new job and found one at the University of Padua, where Copernicus had studied medicine ninety years earlier. The Venetian Republic, the source of funding for Padua, was liberal and tolerant compared with Tuscany, and Padua and Venice were a much more sympathetic venue for someone with radical ideas and an imprudent mouth.

Galileo Galilei, in an oil paint-
ing by Sustermans. *(Corbis)*

Though his financial difficulties didn't end, Galileo was happy
and influential in the intellectual milieu of Padua and Venice. On
the personal front, he fathered three children by Marina Gamba,
though he never married her. Their relationship seems to have
ended amicably when he finally moved back to Florence, for Galileo
remained good friends with her and with the man she later married.
Galileo's two daughters both became nuns. One of them, Sister
Maria Celeste, was a source of great comfort and help to him when
he was elderly, though she died earlier than he, in 1634. The other
two children were not as close to him.

At Padua, Galileo still taught Ptolemaic astronomy. However,
by the mid-1590s it's clear that he was personally convinced that this
astronomy was incorrect. In 1597 (twelve years before he first looked
through a telescope) he wrote a letter reflecting that shift of alle-

giance to a friend at Pisa, and it was also at this time that the exchange of letters with Kepler took place concerning Kepler's book *Mysterium*. Galileo had joined the Copernican camp, but he declined to say so publicly.

In the late spring of 1609, the same year that Kepler published *Astronomia Nova* with his first and second laws, Galileo, in Padua, heard of an invention from Holland currently on view nearby in Venice—a tube with lenses arranged in it so that it made objects in the distance look closer. Apparently Galileo didn't hurry off to examine this wonder in person. A few days later he heard a report of another such instrument from a friend in Paris. His interest aroused, Galileo started pondering what arrangement of lenses would produce the reputed effect. Venice and its nearby islands were centers of expert glass making, so there was no difficulty obtaining the lenses he needed to build his own improved "perspicillum."

The use of lenses for eyeglasses had begun as early as the thirteenth century. There probably were also telescopic devices before the seventeenth century. However, one of the first pieces of documentable evidence of such an instrument identifies the maker as Jän Lippershey, a lens-grinder from the Dutch island of Walcheren. He presented it to Dutch authorities in October of 1608, eight or nine months before it came to Galileo's attention.

Obviously such an instrument would be useful for sighting ships and distant features of the landscape. One brochure in the autumn of 1608 also pointed out its advantage for "seeing stars which are not ordinarily in view, because of their smallness." Sir William Lower, a Welshman, looked at the Moon through a telescope earlier than Galileo did and thought it looked similar to a tart: "here some bright stuff, there some dark, and so confusedly all over."

Clearly, the familiar story that Galileo invented the telescope is untrue. By the time he knew of its existence it had already been on sale in Paris and probably elsewhere for several months. A second piece of fiction is that he tried to pass it off as his own invention. However, Galileo did proceed immediately to make better capital out of it than anyone else was doing. On this occasion, and perhaps several others (it isn't always clear precisely where Galileo got his ideas), Galileo displayed a talent for seeing the unrealized potential of another person's thought or invention and carrying it forward so rapidly and enthusiastically that he was halfway over the horizon before its originator had left the starting line. That isn't plagiarism, but, in the case of the telescope at least, it did result in Galileo getting the popular credit, while his letters show that he was actually giving fair credit to others.

Galileo, a master of self-promotion, presented his own improved version of the tube with lenses to the senate in Venice, hustled some of the senators up to the top of the Campanile, and showed them that it was possible to look out to sea and spot ships that wouldn't be visible to the naked eye until two hours later. The military and commercial advantages of such an instrument were obvious to rulers of a major city-state whose prosperity rested on trade by sea. Galileo received a permanent appointment at the University of Padua and a hefty increase in salary.

But Galileo had uses in mind for his perspicillum other than spotting ships, providing curious occasional glimpses of the Moon and stars, and securing university tenure at Padua. He set about making systematic astronomical observations, recording them, and using his fine mathematical skills to draw conclusions about what they meant.

In the autumn of 1609, Galileo turned one of his new instruments on the Moon. Although the ancient Greeks had described the Moon as "earthy, with mountains and valleys," conventional wisdom in Galileo's day had it that the Moon was perfectly smooth and spherical. Both Greek and Hellenistic astronomers and medieval scholars understood that the Moon shines by reflected sunlight, not by its own light. Through his telescope Galileo watched sunrise on the Moon's surface and saw isolated bright dots in the dark portion expand and join with one another. Reminded of what he had observed when sunrise strikes mountain peaks on Earth, he speculated that the separate bright spots must be peaks and ridges, lit first by the Sun's rays before these could penetrate to the lower areas of the Moon's surface. It occurred to him that by studying the shadows of these features he might measure the heights of the peaks and ridges. Galileo arrived at an estimated height of four to five miles. Modern measurement of the particular range of lunar mountains that he studied has them no higher than 18,000 feet. Never mind that discrepancy. The more significant point was that the Moon was not, as many had supposed, smooth.

Aiming his telescope at things more distant than the Moon, Galileo began to make further discoveries. In January of 1610, using an instrument whose lenses he had ground himself with great care, he discovered three pinpricks of light near Jupiter, neatly lined up with the planet. Galileo watched, mystified and then increasingly excited, as the little stars and Jupiter exchanged positions in their lineup and varied in brightness over the course of several nights. (See figure 3.7.)

Galileo concluded that this remarkable heavenly quadrille "ought to be observed henceforward with more attention and precision." Before long he found that there are four rather than three stars; that the stars move within a narrow range, always in line with Jupiter and with one another; that they stay with the planet when its motion becomes retrograde; that when they are farthest from Jupiter they are never closely packed together, but when they are nearer to Jupiter they are sometimes closely packed. The implication of this last observation was that if the stars are circling Jupiter, the orbits in which they move are not all the same. If the stars were

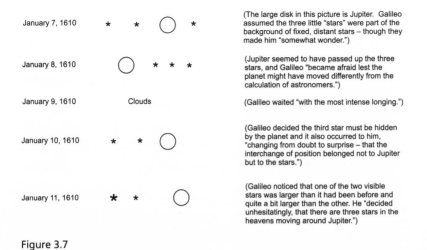

January 7, 1610	★ ★ ◯ ★	(The large disk in this picture is Jupiter. Galileo assumed the three little "stars" were part of the background of fixed, distant stars – though they made him "somewhat wonder.")
January 8, 1610	◯ ★ ★ ★	(Jupiter seemed to have passed up the three stars, and Galileo "became afraid lest the planet might have moved differently from the calculation of astronomers.")
January 9, 1610	Clouds	(Galileo waited "with the most intense longing.")
January 10, 1610	★ ★ ◯	(Galileo decided the third star must be hidden by the planet and it also occurred to him, "changing from doubt to surprise – that the interchange of position belonged not to Jupiter but to the stars.")
January 11, 1610	★ ★ ◯	(Galileo noticed that one of the two visible stars was larger than it had been before and quite a bit larger than the other. He "decided unhesitatingly, that there are three stars in the heavens moving around Jupiter.")

Figure 3.7

following one another in the same track, it's likely they would sometimes line up so as to seem (from our vantage point) to cluster when they are farthest from the star.

Galileo reasoned that these could only be satellites, "planets never seen from the beginning of the world up to our own time," orbiting Jupiter in the same way the Moon orbits the Earth. The deeper importance of what he had discovered also didn't escape Galileo. Never again would it be possible to suppose that there was only one body that was the center of all motion in the universe.

Galileo, being Galileo, soon found a way to capitalize on his discovery. He rushed into print with a book called *Sidereus Nuncius*, translated as *Starry Message* (the quotations in figure 3.7 come from that book), calling on all astronomers to equip themselves with good instruments and turn them on Jupiter. He dedicated his book not to just any local nobleman but to the powerful Grand Duke Cosimo II de' Medici, of Tuscany, who had once been his pupil. He decided to name his discovery the Cosmican Stars—in honor of Cosimo— but soon thought the better of that. It would sound too much like "cosmic," and the significance of the name would be missed. He settled on the Medicean Stars. There were, after all, four stars and four Medici brothers.

Meanwhile, Galileo hadn't been neglecting other stars and planets. He had found that though his instrument transformed the planets into disks, the stars still looked like points of light. Furthermore, there were astounding numbers of them that had never been seen before. The Milky Way, he discovered, is "nothing else but a collection of innumerable stars. . . . many of them are tolerably large and extremely bright, but the number of smaller ones is quite beyond determination." There was far, far more to the universe than anyone on the face of the Earth had ever supposed. Galileo hastily added some pages in the middle of his book to report these discoveries.

A copy of *Sidereus Nuncius* reached Kepler in Prague. He also heard about the discovery of Jupiter's satellites through a friend named Wackher von Wackenfels. Galileo asked Kepler for his opinion, and the reply came in the form of a long letter that was later published as *Conversation with the Starry Messenger*. In it, Kepler discussed Galileo's discoveries and theories and expressed his agreement. Galileo wrote back, "I thank you because you were the first one, and practically the only one, to have complete faith in my assertions." Galileo did not, however, respond to Kepler's rather broad hint that he would enjoy owning one of Galileo's telescopes, though Galileo was sending them as gifts to many influential people. Galileo actually understood the principles of the telescope no better than Kepler, perhaps not as well, though Kepler never built one. When Kepler was working with Tycho's data, he had to study optics to learn how to eliminate errors due to the smearing (refraction) of light as it passes through the Earth's atmosphere. Kepler did, for a short while, have one of Galileo's telescopes on loan from a mutual acquaintance.

Galileo's self-marketing scheme was successful. His book appeared in March 1610—quick publishing indeed—and by late summer he had accepted the grand duke's offer of a job and moved back to Florence. He was now a celebrated and well-placed scientist and astronomer.

About the time Galileo must have been unpacking his equipment in Florence, the planet Venus came into a good position for viewing in the evening sky. Galileo examined the planet and the area around it, searching in vain for companions like the ones he had found around Jupiter. In a letter in mid-November to Cosimo's brother Giuliano, ambassador in Prague, he wrote that there seemed to be no satellites around any of the planets except Jupiter. However, his study of Venus was to yield other significant results. Although most scholars give the credit to Galileo, there is some question whether it was he or his former student Benedetto Castelli who at this juncture remembered a suggestion that Copernicus had made in De Revolutionibus, that Venus might supply important evidence in the case against an Earth-centered universe. It was certainly Galileo who proceeded to find the evidence.

Recall the analogy in chapter 2, with the moving lights on the carousel. Imagine once again the entire amusement park plunged into darkness, with glowing lights attached on the heads of only a few of the riders. As you try to figure out what carnival rides might produce this pattern of movement—and what the park would look like in daylight—bear in mind that some of the pinpoints of light you see might not be light sources at all, but instead be shining by reflected light. Perhaps a bauble that is not a light source itself, on the head of one of the carousel horses, is reflecting the light cast from a nearby horseman. How would you know the difference?

Do any of the lights have "phases" like the Moon? Do any of them sometimes appear as a distorted disk, a half disk, or a crescent? If so, might it indicate that, like the Moon, this is not a light source but a reflection of light coming from elsewhere? Perhaps its "waxing and waning" is a clue to its position and motion, and to the position and motion of the source of its light.

It was this line of reasoning that Galileo used in 1610, when he studied the planet Venus through his telescope. In Ptolemaic astronomy, Venus always lay between the Earth and the Sun. For that reason, *if* Venus sheds no light of its own but only shines with reflected sunlight, observers on Earth should never see the face of Venus anywhere near fully lit. In other words, it should never be the near equivalent of a full Moon. (See figure 3.8.)

From August until October of 1610, Venus would have appeared as a blurry disk through Galileo's telescope. In October he would have seen the disk flatten to a lozenge. Galileo knew then that Venus was shining by reflected light from the Sun, not by its own light. From November until January, Venus waned to a crescent in the same manner the Moon does. Galileo was aware that in the Ptolemaic system it would have been impossible for Venus to have nearly a *full range* of phases, even if its epicycle had been miscalculated and was actually on the other side of the Sun from Earth. To put it bluntly, Galileo could not have seen what he saw if Ptolemaic astronomy had been correct. However, it was what one would *expect* to see in the Copernican system or in Tycho Brahe's system. Finally, Galileo had found persuasive observational evidence that Ptolemaic astronomy was inferior to Copernican astronomy.

To Galileo's mind this was actually not the first such evidence he had found. Six years earlier, in 1604, he had first sided publicly with Copernican theory, announcing in a series of lectures that the nova seen that year, later known as Kepler's Star because Kepler wrote a book about it, provided evidence that some of Ptolemy's arguments were invalid. Since no texts of his lectures survive, it's a mystery what Galileo thought that evidence was. The phases of Venus are a different matter. Clearly, this discovery was a serious setback for Ptolemaic astronomy.

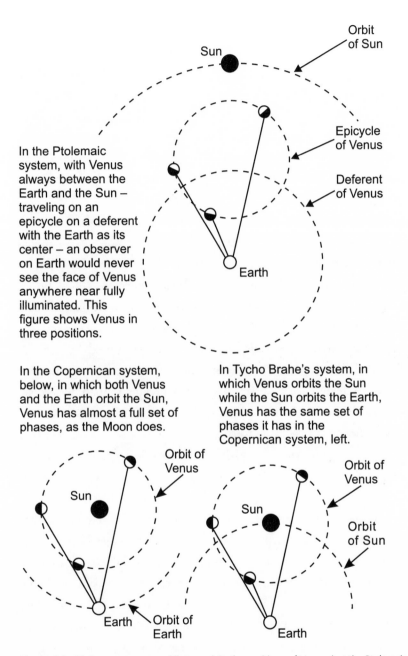

In the Ptolemaic system, with Venus always between the Earth and the Sun – traveling on an epicycle on a deferent with the Earth as its center – an observer on Earth would never see the face of Venus anywhere near fully illuminated. This figure shows Venus in three positions.

In the Copernican system, below, in which both Venus and the Earth orbit the Sun, Venus has almost a full set of phases, as the Moon does.

In Tycho Brahe's system, in which Venus orbits the Sun while the Sun orbits the Earth, Venus has the same set of phases it has in the Copernican system, left.

Figure 3.8 All three systems are able to explain the positions of Venus, but the Ptolemaic system cannot explain the phases. The distances and size of orbits in this drawing do not reflect the actual distances and orbits.

"Eppur Si Muove"

When Galileo published *Sidereus Nuncius* in 1610, though most Catholic believers undoubtedly thought the Earth was the center of the universe and assumed that Scripture supported this view, the Catholic Church had no official policy regarding the arrangement of the cosmos. Contrary to modern popular legend, it had succeeded in staying clear of the debate ever since the appearance of *De Revolutionibus*. In fact, for at least three centuries, it had tolerated what might easily have been seen as nonconformity among its members when it came to cosmological and scientific matters. Nicholas of Cusa had proposed a moving Earth before Copernicus. The Church didn't criticize or condemn him. Giordano Bruno, a scholar strongly influenced by Cusa, was burned at the stake for his views on the Trinity and numerous other heresies. He was rumored to have declared Christ a rogue, all monks asses, Church doctrines asinine, and his own interpretation of ancient Egyptian religion much preferable. Where Bruno thought the center of the universe was hardly mattered when this case came to trial!

Now, in the wake of Galileo's discoveries, many among the Catholic hierarchy appear to have been hoping the Church could continue to avoid getting involved in the controversy, and some among them were particularly anxious that it not take a fundamentalist stand on the interpretation of scriptural passages having to do with cosmology. The quote often attributed to Galileo—"Scripture teaches how to go to heaven, not how the heavens go"—actually was not from him but from one Cardinal Baronis. The vague state of truce seemed to be that if all parties could avoid saying that any scientific arrangement of the universe *or* scriptural cosmological statements should, *or should not*, be treated as literal truth, everything would continue to go smoothly. That state of truce made a great deal of sense at a time when there was no observational evidence to support either theory as a truth claim. The question was: Could it last in the face of what was looking more and more like compelling observational evidence?

As news of Galileo's discoveries spread, the most outspoken and dogmatic reactions came not from churchmen but from conservative astronomers in the universities, who continued to insist that the authority of Aristotle and Ptolemy must not be questioned. Why bother to observe nature or look through a telescope when the ancients had already given the answers? Galileo's battle with these men was not only over the arrangement of the universe. His way of doing science, his insistence on examining and testing nature to learn about it, was to these intransigent scholars foolishness at best, scientific heresy at worst.

However, not all who opposed Galileo were so lacking in valid arguments. Even modern astronomers and historians of science admit that Galileo's case for Copernican astronomy was not as open-and-shut as he insisted it was. Some of Galileo's contemporaries argued, correctly, that Galileo's "evidence" was not "proof." Picking up on what many thought was Copernicus's preface to *De Revolutionibus*, some were willing to accept the Copernican arrangement as an excellent hypothetical model that saved the appearances, while not making decisions on whether it actually represented reality. Jesuit scholars pointed out that while Copernican theory was capable of explaining Galileo's discoveries, Tycho's theory could explain them equally as well without changing the center of the universe.

Galileo himself did not have a great many personal supporters, and his scientific views were not the only cause. He was arrogant, careless of fragile egos, and didn't suffer even the most reasoned opposition graciously. In fact he had a tendency to regard anyone who disagreed with him as an enemy. Vitriolic, insulting statements

in letters and in person didn't make him popular among his colleagues.

Galileo actually bears a large share of the blame for making the
Church the arena where the most visible clash between the two
theories took place. Remarkably, given the Galileo "legend," he
didn't at all oppose the idea that the Church should exercise authority in scientific matters. Quite the contrary, he was impatient and
scornful of its reluctance to face up to the issue. Modern scholars
would consider it healthy that the Church had no official position
about the design of the universe. Galileo did not. What was important to him was that Church authority weigh in on *his* side. He was
supremely confident that it would.

Though Church officialdom continued to be stubborn about
involving itself publicly in this squabble, behind closed doors in the
upper echelons of power it seems certain that the question was not
being ignored and that it was considered a delicate matter. Several
intellectuals among the Church leaders had concluded that Sun-
centered astronomy was preferable to Earth-centered astronomy. If
it was, the Holy Office would be wise to support it. Some thought
it safest to continue the hands-off policy and let matters drift in
that direction, if they did, on their own. Nearly all agreed that the
Church should proceed slowly and with great caution. It had long
felt a duty to protect its members, particularly the less well-educated
among them, from ideas that threatened to undermine simple faith,
however much that simple faith might seem *too* simple to more
sophisticated Catholics. The belief systems of most parishioners
took for granted the Aristotelian/Ptolemaic arrangement of the universe. For this reason a gradual shift was preferable to any sudden
move. It was much less likely to shock and confuse. But also—quite
apart from any genuine concern the Church leadership might have
had for the spiritual well-being of other Catholics—given the tenor
of the times and the ongoing struggle with Protestantism, these

leaders arguably had something to fear from any major unsettling of the flock.

In this setting, Galileo was the proverbial bull in the china shop. He had too much zeal for his cause to recognize that forcing the issue might not be in his own best interest. He was impatient. He knew that some powerful men were agreeing with him. Surely others could not fail to see that on both scientific and scriptural grounds, Copernican theory was correct and acceptable.

In 1611, Galileo paid a visit to Rome and was received with enormous respect and friendliness. But two years later he made a serious error. He wrote letters to his former student Castelli and to the mother of the duke of Tuscany, stating that Copernicanism should be treated as fact, not hypothesis, and commenting that it would be unwise to interpret passages of Scripture in such a way as to force them to support interpretations of nature that might later prove obviously wrong. (It seems Galileo thought that only a "forced" interpretation of Scripture could support Ptolemaic astronomy.) Some in high Church circles had been voicing similar sentiments, recognizing that this was not a contest between literal and metaphoric reading but a matter of *which* metaphoric reading. Unfortunately, Galileo also said in his letters that Scripture had not gone into more sophisticated scientific detail which *could* be taken literally because it was written to be understood by "common people who are rude and ignorant."

Galileo's letters got passed around, and soon some were interpreting them (or deliberately misinterpreting them) to say that Galileo was questioning the validity of Scripture. When pressed, a committee of the Holy Office looked into the matter and ruled that Galileo's letters were *not* heretical. However, he received a letter from the powerful Jesuit Cardinal Roberto Bellarmine, advising him that until there was definite proof that the Earth moved, that idea ought to be treated as hypothetical and Scripture interpreted in the more commonly accepted way. Bellarmine was one of those who believed that was how Copernicus himself had regarded his model.

In 1616, Galileo was once again in Rome pushing his campaign for Copernicanism. This time the Holy Office, increasingly belea-guered by Galileo's enemies, sent a survey of sorts to leading univer-sity scholars regarding the centrality of the Sun and the motion of the Earth. It isn't known who drew up the mailing list, but the replies were overwhelmingly anti-Galileo: He was scientifically wrong and philosophically heretical.

Galileo's opponents won a qualified victory in this round. Gio-vanfrancesco Buonamici, a diplomat from Tuscany, recorded in his diary that two cardinals opposed the pope's inclination to declare Copernicanism contrary to the faith. Bellarmine was one of them. The other was Cardinal Maffeo Barberini, a good friend of Galileo's. In the end the judgment was that Copernicanism was "contrary to Holy Scripture and cannot be defended or held." However, that judgment was not given the stamp of papal authority, and this meant that Copernicanism was not officially heresy. That seems a trivial legal distinction, as does another, that Galileo had been in-formed of the Church's decision and admonished to abandon Coperni-can views until he had unassailable proof, but not punished or forbidden to teach Copernicanism. However, Galileo recognized this last point as so crucial that he asked for and received a letter from Bellarmine, who had delivered the Church's verdict to him, making the details clear. Galileo did stop campaigning for Coperni-can theory for a while and turned to other work, biding his time and watching for a better political moment to resume that effort.

In 1623, that moment seemed to have arrived. Liberal Catholics rejoiced as Cardinal Barberini became the new Pope Urban VIII. Barberini and Galileo had feasted often together and enjoyed dis-cussing science and philosophy. Barberini, as a cardinal, had suc-cessfully opposed the papal decree that would have declared

Copernicanism heretical. Some accounts have it that now, in 1623, Barberini told Galileo that as pope he could no longer chat off the record with Galileo and express his own opinions freely. Galileo asked to be allowed to write a book on *both* systems, and he took Barberini's response, whatever that actually was, as encouragement.

By this time, in spite of the 1616 stalemate in Italy, Copernican astronomy—with Kepler's elliptical orbits and his three laws, Galileo's discoveries, and the *Rudolphine Tables* soon to be published—was very near to holding its winning hand. But if Galileo thought the Church was ready to put its cards on the table and support Sun-centered astronomy, he was mistaken.

The story gets increasingly murky. Perhaps Barberini was doing his best to be both a cautious Church ruler and a friend to this brilliant, volatile man. Perhaps he still hoped to let matters evolve more gradually. A more politically sensitive and diplomatic person than Galileo might have recognized this and carried through on the task of writing an impartial book, one presenting both theories and describing his own observations without highlighting their fullest implications. Such a book might have achieved a significant victory for Copernicanism, without requiring him to become a martyr in the process.

Galileo was almost surely not thinking of martyrdom as a possible outcome. What *was* he thinking? Did he decide, remembering the Church's earlier policy of tolerance, that the 1616 judgment had been only a temporary conservative aberration? Did he take Barberini's statement that as pope he could no longer chat "off the record" to mean that he couldn't tell Galileo outright to author a pro-Copernican book but would like to see it happen? Or did Galileo, convinced as he was of the irresistible nature of his arguments and evidence, just get carried away with missionary zeal and the force of his ideas as he wrote the book? Significantly, he now believed he had discovered the "proof" Bellarmine had challenged him to find.

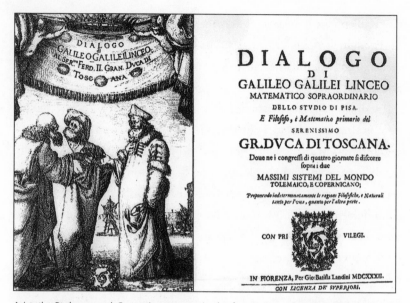

Aristotle, Ptolemy, and Copernicus appear in the frontispiece of Galileo's *Dialogo*, published in 1632. The title page assures readers that this "discourse concerning the two chief systems of the world, Ptolemaic and Copernican," discusses "without prejudice one view, then the other, on the basis of philosophy and natural law." *(Frontispiece and title page,* Dialogo, Galileo Galilei*)*

In any case, Galileo's book, though it was to seal the triumph of Copernican astronomy in the long run, in the short run played directly into the hands of his enemies. *Dialogo . . . sopra i due Massimi Sistemi Del Mondo* (Dialogue Concerning the Two Chief World Systems), more often called by its shortened Italian title, *Dialogo*, makes a powerful case for Copernicus. Unlike *De Revolutionibus*, it is not a technical, mathematical book. It is an entertaining masterpiece of popularization, written in Italian rather than in scholarly Latin, designed to appeal to a great many readers—as no book on astronomy ever had before. It takes the form of a lively four-day discussion among three friends. The first day is used to demolish the ideas of Aristotle. The second and third days are devoted to proving that the Earth turns on its axis and orbits the Sun. The fourth is spent on Galileo's new theory of the tides, which is in error but which he thought was his clinching argument for Copernican astronomy. (Galileo in fact called Kepler childish for supposing that the Moon affects the tides.)

Galileo left no doubt whatsoever that the character Salviati, who is a far more intelligent man than either of the other characters and who argues eloquently for Copernican theory, represents Galileo himself. Salviati has the floor much of the time. The character who argues the Aristotelian/Ptolemaic side of the case—Simplicio—gets the last word, but he is little better than a buffoon, confused and slow. One is tempted to say that he provides the comic interest. Lest any of Galileo's readers miss the point, a third character, Sagredo, a skillful discussion leader who asks thoughtful questions to move the conversation along, scolds Simplicio for his stupidity and inability to see reason and congratulates him condescendingly whenever he does see the light just a little. Galileo was a master of ridicule, and he used that weapon mercilessly in *Dialogo*. Doubtless it seemed to him that anyone who would continue to defend Aristotle and Ptolemy in the face of compelling arguments for Copernican astronomy *could* only be a half-wit. Galileo surely could not possibly have thought his book was impartial or that anyone else would think it was.

Amazingly, there was initially no adverse reaction! The book made its way through Church bureaucracy with only a few annoying delays, which only with hindsight look ominous. After minor changes, it actually won approval from the Church censors, who made one inadvertent improvement by requiring Galileo to change the title from *Dialogue Concerning the Tides* to *Dialogue Concerning the Two Chief World Systems*. The censors had no way of knowing that the section of the book about the tides was its one embarrassing weakness.

It might begin to seem at this point that Galileo had judged the situation entirely correctly. *Dialogo* appeared in February 1632 with a great flurry of publicity, and the reception was overwhelmingly enthusiastic. There was, to be sure, some criticism, notably from one Christoph Scheiner, a Jesuit who had earlier clashed with Galileo about which of them had first discovered and correctly interpreted sunspots. Galileo had pretty well demolished Scheiner in that encounter and was not gracious about the victory. It comes as no great surprise that Scheiner was more than a little sour about Galileo's latest triumph.

Then, out of the blue, disaster struck. Barberini, Pope Urban VIII, Galileo's old friend, suddenly turned bitterly against him. There is no documented reason for Barberini's belated change of heart. The argument that the pope was under political and military pressures that caused him to change his mind about Galileo is unconvincing. These problems were so far removed from Galileo and the Copernican question that it is difficult to imagine they had an impact except to contribute to a state of tension. It also makes little sense to say that the Church could not seem to waver on any decision. On what decision was it wavering? The 1616 dictum? Copernicanism had not been officially declared heresy then, and lest anyone had missed that technicality, much more recently Church censors had approved Galileo's book. Why waver from *that* decision? Barberini seems to have unleashed the fury of the Church against Galileo *in spite* of the fact that he must have known the result would be to make the Church and himself look vacillating and foolish. Many devout liberal Catholics were shocked when the Church finally condemned and prohibited the teaching of Copernicanism, because it committed the Church to a theory that now looked untenable. Barberini surely anticipated this reaction.

One popular explanation is that advisers around the pope convinced him Galileo had meant Simplicio to be a caricature of Barberini himself. Galileo had made the mistake of putting in

Simplicio's mouth an argument that Barberini had once expressed in their discussions. Even so, the reaction seems extreme. The interpretation that makes this episode into a major confrontation between science and religion is that the pope saw Galileo and Copernicanism as a threat to belief in the validity of Scripture and to the Church's right to be the final arbiter of truth. In fact, such motivation might have pushed a good politician, which Barberini was, to side with the theory that looked likely to win the day. Approving Galileo's book was a strong move for the Church. A more likely reason for Barberini's ire is more subtle—that Galileo had finally managed to make it appear to the public that the Church had thrown its weight solidly behind Copernican astronomy. The decision to do that should have come, if it came at all, from Church officials, not from Galileo. The issue had little to do with Church/science authority, but *everything* to do with Barberini/Galileo authority. Barberini might indeed have felt he had been betrayed and used for a fool and must now take desperate damage-control measures.

When the printer of *Dialogo* received the surprising order to send all unsold copies of the book to Rome, he couldn't comply. They were sold out. But the news soon got about that *Dialogo* was to be reexamined to determine whether it was heretical. The Holy Office sent for Galileo in February of 1633. He went to Rome and resided at the home of the Tuscan ambassador for some weeks while the Holy Office scrambled to put together a coherent case—no easy matter, since Church censors had found no fault with Galileo's book and the opinion that Copernican theory was heresy had never been the official opinion of the Church. Church legal experts nevertheless did their best (or worst), and Galileo stood trial for "a vehement suspicion of heresy." Galileo's letter from Cardinal Bellarmine that said Galileo had not been forbidden by the 1616 decree to teach Copernicanism carried little weight, though it did cause chagrin.

Even less effective was Galileo's pitiful defense strategy. He told the court that his accusers had misinterpreted *Dialogo*. The book was actually a defense of Ptolemy, and he could add some pages at the end to make that clear. Even the least astute among the inquisitors could not be expected to believe this, for Galileo had, in *Dialogo*, called defenders of Ptolemy "dumb idiots."

Galileo was not tortured, nor was he sentenced to death. It seemed to be only his complete humiliation that the pope wanted. Galileo was forced to renounce Copernicanism in a long, demeaning statement before a resplendent crowd of dignitaries, nearly all of whom must have known the old man didn't mean a word of it. The story goes that he made his exit muttering "eppur si muove" (and yet it does move). Whether or not he really said the words out loud, he surely was thinking them.

Galileo was sentenced to spend the rest of his life in isolation, under house arrest at his own villa at Arcetri. Some scholars argue that the only thing that saved him from a harsher punishment was that Barberini, for all his rage, had enough foresight to realize that no matter what the court declared, Galileo had won—that history and many of Barberini's contemporaries within Church officialdom would ultimately condemn Barberini for upholding Earth-centered astronomy. As it turns out, history remembers him as a cruel, irrational bigot. His Church and Catholic scholarship and science suffered irreparable damage and loss of credibility. After the trial, Italy lapsed into what was almost a scientific dark age with the prohibition of Copernican theory. The center of scientific endeavor and achievement shifted to northern Europe and England, never to return.

Galileo lived for eight years after his trial in 1633. He was in his seventies, but still active. It was during these years that he got around to publishing much of the scientific work that he had carried out when he lived in Padua. Many scholars consider his *Discourse on Two New Sciences*, produced during these years and again starring

Salviati, Sagredo, and Simplicio, as his greatest book. It doesn't deal with the Copernican question. Galileo eventually went blind, and he died in 1642 at the age of seventy-eight.

Did Galileo know that though he had suffered personal defeat he had won the war for Copernicanism? Probably. It's less probable he realized that he would go down in history as a scientific martyr and a symbolic figure for all those who see religion as the enemy of science, and vice versa. He would not have liked that, surely. For he was so firmly convinced, and argued so well, himself, that there was no contradiction between good science and belief in God.

Weighed in the Balance

As Galileo, Kepler, and other sixteenth- and seventeenth-century scholars made up their minds whether or not to become Copernicans, some less-than-familiar scientific values came into play. It certainly wasn't always just a question of what fit best with observation—not even with Galileo, who has been dubbed the father of modern science. Why else prefer one model to another?

Suppose that you and I, for reasons we won't pause to examine, propose that the Moon is the center of the system. We know that the issue here is one of relative motion only, and that we can come up with a Moon-centered model that does indeed fit with all the data and that no one can prove is wrong. What, then, *would* make anyone prefer another model to ours? What has Copernicus got that we haven't?

Imagine our planetary system represented as dots moving on a three-dimensional supercomputer screen. Freeze first one dot and then another, each time allowing the computer to make sense of the movement of the other dots in terms of the unmoving dot. The picture may always be correct and accurate, but it will sometimes

look far more complicated, and sometimes far simpler. The choice
of one particular dot, the Sun, as "center" causes the picture to
fall into place and appear remarkably simple and harmonious. Why
bother with the other ways, just because they don't happen to be
"wrong"? This reasoning causes scientists to prefer the Copernican
model to Tycho Brahe's.

Simplicity and harmony are strong pointers for scientists but
not absolute clinchers of the sort Mr. Elmendorf wanted. There are
other criteria. Modern science backs up a choice of *how* the solar
system moves with the answer to a second question: *Why?* It isn't
enough to claim that a model can predict where all the planets will
be a month or a year or a century from now. What makes them go
there? To use more scientific language, what are the "dynamics" that
cause them to move in this way rather than another? What makes
them move at all rather than sit still? There were those who thought
that the Creator had given an initial push to each planet and that
that movement would go on forever. Others suggested that angels
move the spheres. Ptolemaic astronomy explained the movement as
originating in the sphere of the stars and then being transferred to
the planetary spheres. Copernicus didn't attempt to explain why the
planets moved as he thought they did, although he recognized the
importance of the question. Kepler believed a whirling force ema-
nating from the Sun drives the planets and that this could only
work if the Sun were in the center. But not until Isaac Newton
and later Albert Einstein, with their explanations of gravity and
the curvature of space-time, was anyone able to suggest the reasons
accepted today why the planets move as they do. Is that how scientists know Copernicus was right? Not quite. Fred Hoyle has argued
that a subtler understanding of Einstein's theories reveals they may
actually slightly favor an Earth-centered model. Had Galileo had
Hoyle at his elbow, he might have produced the book that would
have pleased the pope and not have been tried for heresy!

Why, then, does Ptolemy come off so badly in this contest? Paradoxically, the enormous success of Ptolemaic astronomy is not an argument in its favor. It can account for all apparent movement in the heavens. It could also account for a great deal that never happens. It allows for too much. Copernican astronomy, as it has evolved, allows for far less. It's easier to think of something that Copernican theory could not explain. The more scientific way of putting this is that Copernican theory is more easily "falsifiable" than Ptolemy's, easier to *dis*prove. Falsifiability is considered a strength. If a theory sets up a clear enough profile so that it provides numerous opportunities to shoot it down, and no one is able to shoot it down—if new discoveries don't undermine it but fall neatly into place, as Galileo's discoveries with his telescope did—if the picture gets more harmonious over time rather than more complicated—then a theory begins to look as though it is on target.

There is another criterion by which theories are judged, and, for better or worse, it shows that modern scientists do have a certain kinship with those recalcitrant seventeenth-century scholars they so disdain. When new theories and the implications of new discoveries disagree with the way a scientist personally feels the universe ought to run, he or she is reluctant to accept them. "It can't be right, because it feels so wrong." In the twentieth century, Einstein's resistance to the notion of an expanding universe, as well as many scientists' discomfort with the possibility that the universe may be less predictable than previously thought, are examples. One need not be narrow-minded or antiscience to dismiss a theory on the grounds that it offends one's scientific aesthetic sensibility.

These few paragraphs have viewed the competition with the eyes of twentieth-century philosophy of science, at the end of a century that has given an extraordinary amount of thought to the question of how we know what we know. But modern thought processes are rooted in thinking that was already substantially in place even in the ancient world and certainly in Kepler's and Galileo's time. Scholars didn't think in terms of relative motion in the twentieth-century sense, but they knew that there could be competing and equally valid geometric explanations. When it came to making simplicity a criterion, scholars had no "scientific method" yet to tell them that the most economical and harmonious model that saves the appearances is the best model. But even the ancient astronomer Hipparchus was known as a man who always preferred the least complicated hypothesis compatible with observation, and Neoplatonism brought into late medieval science a preference for finding simple mathematical and geometric regularities in nature. Copernicus firmly believed in the superior harmony of his arrangement and saw that as a powerful argument in its favor. The search for such harmony was the driving force behind Kepler's work. Some scholars expressed this preference in religious terms: The work of God and nature is work of great elegance and simplicity, and we must try to understand and explain similarly.

Scientists, philosophers of science, and historians find it no easy matter to fathom precisely what causes one theory to emerge as "scientific knowledge" while another is discarded. Their explanations almost always involve personalities, history, and happenstance as well as science. The stories of Ptolemy, Copernicus, Kepler, Tycho Brahe, and Galileo discourage simplistic interpretations in terms of proved/unproved, bad science/good science, knowledge/ignorance, pro-Ptolemy/pro-Copernicus, religion/science, open-mindedness/ blind dogmatism. Indeed, trying to unravel the complex human saga of the Copernican revolution makes explaining the night sky, by comparison, seem a relatively simple, straightforward task.

An Orbit
with a View

{1630–1900}

> ❝What a wonderful and amazing scheme have we here of
> the magnificent vastness of the universe! ❞
> **Christiaan Huygens**

When French astronomer Pierre Gassendi watched the transit of
Mercury in 1631 by projecting the Sun's image on a circle divided
into degrees, he couldn't believe what he saw. The tiny dot on the
Sun must surely be a sunspot, not the planet. Gassendi, along with
the rest of his generation, had studied Ptolemy's estimates of cosmic
dimensions, and Kepler himself had also predicted a much larger
apparent diameter for Mercury than what Gassendi was seeing.
Only when Gassendi realized that the dot was moving much too
quickly across the face of the Sun to be a sunspot did he conclude
with astonishment that, in spite of its "entirely paradoxical smallness," this was indeed Mercury. That transit permitted the first accurate measurement of Mercury's apparent size. (Apparent size
means the size of the image of an object in the sky as it appears to
us—not its true dimensions, which can't be known without also

knowing its distance.) A Venus transit the same year, also predicted by Kepler, was not visible from Europe, but eight years later there was an opportunity to measure the apparent diameter of that planet as well.

Ptolemy's cosmic dimension estimates had been a speculative offshoot of his mathematical astronomy. He had even commented in the *Almagest* that the absence of observed transits of Venus and Mercury might be because of their apparent smallness, due to great distance. But it's clear from both scholarly writing and popular literature that the Middle Ages had forgotten such quibbles and regarded his estimates as established truth, and Ptolemy's numbers still had a tenacious hold on the minds of scholars in the early seventeenth century. However, in the 1630s, when astronomers like Gassendi began using telescopes on a day-to-day basis and became familiar with Kepler's laws, faith in the ancient measurements began to crumble, at first among specialists and then among all educated people. Ptolemy's distance estimates were far too small.

The quality of telescopes was improving, and there were also advances in related technology. In the spring of 1610, Galileo had been using a telescope that magnified thirty times. Two years later, astronomer Thomas Harriot had one that magnified fifty times, but at this magnification the area visible through it was frustratingly small. Then in the 1630s, an instrument known as the astronomical telescope came into use. Kepler had first introduced the theory of this telescope in 1611, but his version inverted the image. When changes in the configuration of lenses solved this problem, the astronomical telescope came into its own, providing a much larger field of view at high magnifications.

The astronomical telescope had another great advantage. Yorkshireman William Gascoigne discovered it by accident in the late 1630s when a spider spun a web in his telescope and he saw some of the threads of the web sharply outlined against the background image. What this told Gascoigne was that an object inside an astronomical telescope could appear in sharp focus, superimposed on a distant object under observation. In other words, he could place some sort of ruler inside his telescope. The ruler he built was a micrometer (as the name suggests, an instrument for measuring extremely small dimensions), with cross-wires that moved across the image with the turning of a screw. Before this invention it was possible only to *estimate* the apparent size of a body. Now there was a way to *measure* the apparent size, comparing it directly with an established standard. Sadly, Gascoigne died in the battle of Marston Moor in 1644, and his invention remained unknown to other astronomers until the 1660s. By then, French experts had developed a similar idea from astronomer Christiaan Huygens. The Royal Society in England was quick to claim Gascoigne's prior discovery, but Frenchmen Adrien Auzout and Jean Picard were largely responsible for developing the screw micrometer into a fine precision instrument.

By 1675, observers were beginning to agree on the apparent sizes of the planets. However, consensus about their distances (and, of course, their true sizes as well), now that Ptolemy was no longer to be trusted, was slower in coming. Kepler's third law of planetary motion had established a relationship among the planets having to do with their orbital periods and their distances from the Sun, but Earth-bound observers were still little better off in terms of absolute measurements than if they had been standing at the foot of a ladder leading up into the sky, knowing that the second rung was twice as far as the first, and the third three times as far, and so forth, but having no clue how far up the first rung was.

The Enormous Advantage of Being in Two Places at Once

High up in the southern wall of the Bologna Cathedral of San Petronio, there is a small opening through which a shaft of sunlight penetrates to the cathedral floor below. There, a gnomon—a protrusion like the raised part of a sundial—makes possible the measurement of the shifting image of the Sun. In the mid-seventeenth century, alterations to the cathedral made the old gnomon useless, and Gian Domenico Cassini of the University of Bologna was given the task of setting up a new one to replace it. Cassini was a good choice, for he was keenly interested in the Sun and its movements. Cassini's gnomon still stands in San Petronio.

In 1669, when Cassini was forty-four, Jean-Baptiste Colbert, an influential minister in the court of King Louis XIV of France and member of the recently founded Academy of Sciences in Paris, invited Cassini to come to Paris to work at the new Royal Observatory. Louis XIV was the king who built Versailles. He cultivated a public image of himself as the Sun King and surrounded his sumptuous court with appropriate symbolism. Naturally, his chief astronomer should be not only one of the best in Europe but also an expert on the Sun.

There were additional, scientific reasons for luring Cassini to Paris. Members of the Academy were already deeply interested in solar theory and eager to gain a better understanding of the Sun, the planets, and the "refraction" of light by the Earth's atmosphere—that is, the way the atmosphere bends and smears light rays passing through it. It was their intention to bring astronomical tables into line with new measurements provided by telescopes and micrometers. Cassini accepted the invitation and became head of the Observatory. The French version of his name was Jean-Dominique Cassini.

All over Europe at this time, telescopes were getting longer and longer. Some experts even voiced the extravagant hope that a long enough telescope would allow astronomers to study the animals on the Moon. One of Cassini's first concerns was to make sure that the new Observatory acquired some of the finest instruments available. In 1671, he brought from Rome a 17-foot telescope manufactured by his good friend and former colleague Giuseppe Campani, one of the most skilled telescope builders in Europe. Colbert, the courtier at whose invitation Cassini had come to Paris, was particularly taken with this telescope, and his enthusiasm prompted the King to present the Observatory with a still larger Campani telescope, measuring 34 feet, and later, in 1684, a 100-foot model. The old wooden Marly water tower was trundled over to the Observatory to serve as a support for these giant instruments, with steps built to the top and a rail added to prevent astronomers and their assistants from toppling off on dark nights. Among the discoveries Cassini made with these telescopes were several moons around Saturn and a dark division of Saturn's rings. His most famous achievement, however, was the first successful measurement of distances within the solar system.

Cassini and other astronomers knew that in August and September of 1672, Mars would be at its nearest proximity to the Earth, and this would provide optimum conditions for measuring its distance. The method Cassini planned to use was an old mapmaker's trick, known as triangulation, in which two observers in two widely separated locations measure the position, against the background, of the same distant object at the same time. Because they are far apart, the two observers have different viewing angles.

Human beings and other animals use this technique instinctively to judge distances. Our two eyes are the two "observers." The view from the right eye is not exactly the same as the view from the left. Hold a finger upright close in front of you, shut one eye, then the other, and the finger appears to change position relative to the background, although it hasn't actually moved. The closer your finger is to your face, the greater the shift appears to be. You don't use geometry to measure the distance from your finger to your face from how great the apparent shift is. Your brain does it automatically with a precision that is adequate for most everyday purposes.

The apparent shift in the position of an object against the background, as seen from two locations, is called a parallax shift. To understand the use of this shift for measuring distances, imagine you are driving along a road across a desert. The desert stretches in all directions to the horizon, where a string of mountains is just barely visible. You notice a solitary cactus not far from the road. You stop your car and ponder the question: How can you measure the distance from the road to the cactus without leaving the road?

Begin by constructing two imaginary triangles: The first triangle has the road as one of its three sides and a line drawn straight to the cactus as a second side. (See figure 4.1a.) The letter X designates the point at which the line to the cactus meets the road. As you stand at X, look toward the cactus and find a landmark directly behind it on the horizon. Fortunately, there is one, a snowcapped peak that stands out clearly from the other mountains. From X, move a little to the left along the road. Again look at the cactus and the mountain peak. They will no longer appear to be lined up. The peak will have moved out from behind the cactus. The location where you have stopped to take a second look is Y on the drawing. (To associate this with the finger exercise above: X is like the view from your right eye, and Y is like the view from your left.)

The next step is to build the second imaginary triangle. One corner of this triangle is the point where you now stand (Y). One side of it is an imaginary line drawn from you at Y through the cactus off to the horizon. There, directly behind the cactus, is an-other convenient landmark, a particularly precipitous crag. You don't have to know the length of the line to the crag. A second side of the triangle is an imaginary line drawn from you at Y to the peak. Again, you don't have to know the length of the line. (See figure 4.1b.) What you do need to know is the "angular distance" between the peak and the crag.

Angular distance is the number of "degrees" between two lines like those pointing from you at Y to the peak and the crag. It works

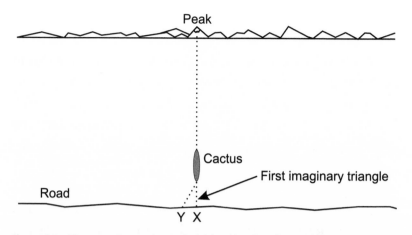

Figure 4.1a The cactus is near the road and there is a ring of mountains on the horizon. From X, on the road, the cactus is directly lined up with a snowcapped peak. Y is a little to the left of X. There is an imaginary triangle whose sides are a line from X to Y, a line from X to the cactus, and a line from Y to the cactus.

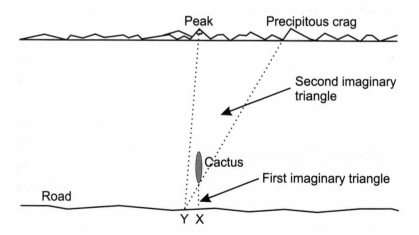

Figure 4.1b Draw lines from Y to the peak and from Y through the cactus to the horizon. The second line is a continuation of the line previously drawn from Y to the cactus, and it leads to a crag on the horizon.

like this: If you are at the center of a circle, and the crag and peak are on the rim of the circle, how many degrees apart on that rim are they? A circle has 360 degrees. A quarter way around a circle (from twelve o'clock to three o'clock) is 90 degrees. (Lines drawn from twelve and three o'clock to the center of a clock meet there at a 90-degree angle.) The angular distance between the peak and the crag is obviously not that great. It's more like the distance from twelve o'clock to one o'clock. That's 30 degrees.

Now pretend you are hovering in a helicopter over this scene, for it's necessary to draw an idealized picture of it from that vantage point. In figure 4.2, the large circle is the horizon, with the mountains. The dot in the center is the cactus. The segment of the road between X and Y is part of a much smaller circle, centered on the cactus and near to it. You don't know the size of this circle, but that doesn't matter. It's drawn as a broken line. The huge circle of the horizon has become a giant clock face, aligned so that the snowcapped peak is at twelve o'clock. The two lines that meet at Y, on the road, end up widely separated at the horizon, at the peak and the precipitous crag. The crag is at one o'clock. Viewed from the center, where the clock hands would be attached, the distance from twelve o'clock to one o'clock represents a shift of 30 degrees. The angle at Y (very close to the center) is *approximately* a 30-degree angle.

So far you may not seem to have found out very much. Actually you have all the information necessary to calculate the distance to the cactus from the road. Taking stock: You can pace off the distance between X and Y on the road—an easily measurable distance. You know the approximate angular separation between the lines leading from Y to the peak and the crag (30 degrees). The rules of geometry say that you, standing at Y, will measure almost exactly the same angular separation between the peak and the crag as the cactus, looking back at the road, would measure between X and Y. You can check this out. On the clock-face drawing, lines drawn

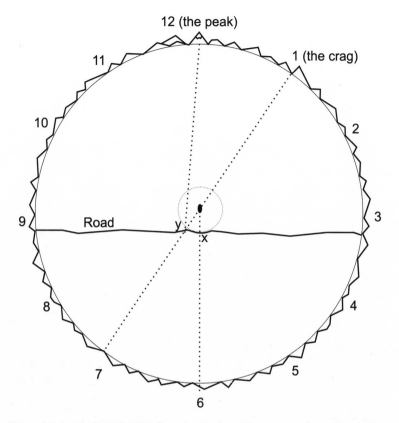

12 (the peak)

11

1 (the crag)

10

2

9   Road   y

x

3

8

4

7

5

6

Figure 4.2 An idealized drawing, from the air, shows the cactus at the center of a huge circle (the horizon), visualized as a clock. The part of the road between X and Y is a segment of a much smaller imaginary circle centered on the cactus. Lines drawn from Y to the peak and through the cactus to the horizon form approximately a 30-degree angle at Y. Also, lines drawn from the cactus through X and Y form a 30-degree angle at the cactus.

from the cactus through *X* and *Y* pass approximately through six o'clock and seven o'clock, and that means the angular separation between those two lines, also, is approximately 30 degrees. If the horizon were as far away from the cactus and you (at *X* or *Y*) as the stars are from Mars and the Earth, the angles would be even closer to identical.

Continuing with the measurement: There is only one distance the cactus could be from the road, at which the cactus would measure a 30-degree angular separation between X and Y, and at which the *actual* distance between X and Y along the road would be the distance you paced off.

To apply the same method to the distance to Mars: The cactus is Mars, and the horizon with the ring of mountains is the distant stars. Astronomers had to find two viewing positions analogous to X and Y; those positions had to be accessible and their distance from one another measurable, and they had to be far enough apart for a parallax shift to be detectable. The distance between the two positions would become the "baseline" for the measurement, a tiny part of an imaginary circle analogous to the small one in figure 4.2. In the Mars measurement, this imaginary circle, with Mars in the center and Earth on the rim of the circle, would be huge viewed from the Earth, but so minuscule compared to the great ring of distant stars that Mars and the Earth could be thought of as both being, essentially, at the center of the star ring—the center of the clock. The parallax shift of Mars as viewed from the two ends of any baseline possible on Earth is not anywhere near as large as 30 degrees.

In order to have observers in two widely separated locations on the face of the Earth whose distance from one another was known, Cassini and his colleagues decided to put into action what had previously been vague plans for the Observatory to send an expedition to the tropics for other astronomical research. Fortuitously, Colbert was hoping to establish a colony at the mouth of the Cayenne River in South America in what is now French Guiana, where there had been French settlers earlier in the century. Ships were sailing there regularly. Cassini dispatched a young colleague, Jean Richer, and an assistant, a Mr. Meurisse, to Cayenne equipped with several measuring devices and telescopic sights. Their instructions were to make observations leading to the measurement of the parallaxes of the Moon, the Sun, Venus, and especially of Mars.

In Cassini's triangle (analogous to the "first imaginary triangle" in figures 4.1a and 4.1b), the three corners were Paris, Cayenne, and Mars. He knew approximately the distance from Paris to Cayenne (the longitude of Cayenne was not certain, but that was something the expedition hoped to remedy), taking into account the curvature of the Earth. That distance was the baseline—Cassini's "distance from X to Y." It was possible, though problematic with the instruments available to Cassini, to detect the parallax of Mars from that baseline. Parallax measurement can, in principle, be made to work for anything that can be seen to have a parallax shift. Figure 4.3 demonstrates (though not to scale or with the angles that really exist) parallax measurements of the Moon and Mars.

Cassini's measurement to Mars was not as simple as finding the distance to a cactus, for several reasons. A cactus isn't going anywhere. A planet is. If Cassini and Richer couldn't time their measurements precisely and know how the time of a measurement in Cayenne compared with the time of a measurement in Paris, the resulting data would be worthless. Cassini couldn't merely give the order "synchronize watches." The best clocks available were pendulums. They could be synchronized, but they wouldn't stay synchronized while one of them was taking a sea voyage to the other side of the world.

During his years at the University of Bologna, Cassini had studied the eclipses of Jupiter's moons and their shadows as they crossed the body of the planet, and he had drawn up tables of their motions. In 1666 he'd noticed that the moons were close enough to Jupiter so that a moon's appearance from behind Jupiter would be seen simultaneously from any point on Earth, and he realized that Jupiter and its moons could provide a way to determine the time difference between widely separated points on the Earth.

Other problems couldn't be so handily solved and caused some later experts to be critical of Cassini's blunt announcement of a definite parallax for Mars, for Cassini was well aware of the inevita-

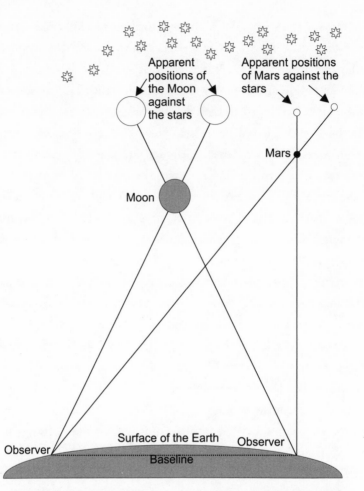

Figure 4.3 Parallax shifts of the Moon and of Mars as their positions are seen from two widely separated points on the surface of the Earth. [Note: This figure is not drawn to scale.]

bility of a large margin of error in his measurements. In fact, probably no one was better able to appreciate that margin of error than he, because of his earlier study of the refraction of light by the Earth's atmosphere and his own previous efforts to measure the Sun's parallax.

Nevertheless, in the summer of 1673, Cassini and the Academy awaited Richer's return from Cayenne with intense excitement.

When Richer arrived in August, Cassini set immediately to work analyzing the data from Cayenne and Paris and additional locations in France where there had been observations. He found much of it not too revealing, but when he had completed his analysis, he reported that from the parallax of Mars he had derived a distance from the Earth to the Sun of 87 million miles (140 million kilometers).

Others besides Cassini took advantage of Mars's unusual proximity. In England, a young, largely self-educated astronomer named John Flamsteed had been studying the Sun, Moon, and planets. He decided that an old method Tycho Brahe had used to measure parallax would work much more successfully now that there were telescopes. Tycho's method didn't require observers at widely separated locations. One observer would suffice to measure a "diurnal parallax," which means a change in position as observed from one single location on the Earth's surface at two different times of day. The rotation of the Earth carries the observer from one end of the baseline to the other. Though Flamsteed's father sent him on a business trip at precisely the wrong time, Flamsteed did manage to observe on one clear night, and he also came up with parallax calculations for Mars. His results were largely in agreement with Cassini's.

Cassini's and Flamsteed's findings did not, as is often supposed, bring immediate consensus among astronomers about the distances to the Sun and the planets. Others besides Cassini were aware of the large margin of error. However, though these measurements were still imprecise and somewhat in dispute, they became the key to the solar system. Now it was possible to use Kepler's laws to calculate distances from the Sun to all the known planets. The Sun was astoundingly far from Earth. Copernicus had calculated the distance as 2 million miles; Tycho Brahe, 5 million; Kepler, no more than 14 million. The new measurement was 87 million miles or 140 million kilometers. (The modern measurement is 93 million miles or 149.5 million kilometers.) In the late 1670s, men and women thus for the first time became aware of the size of the solar system and found it enormous beyond anyone's previous imagining. The universe beyond must be inconceivably vast. In the Middle Ages, the distance to the firmament of stars had been illustrated by how many years Adam, starting his journey on the day of creation, would still have to walk at a rate of twenty-five miles per day to reach them. We hope he brought provisions for a long hike. The new illustration had a cannonball traveling at 600 feet a second (considerably faster than Adam, and presumably passing him as it went) taking 692,000 years.

Cassini went on attempting to measure parallaxes and to tease fresh results out of the data from the Cayenne expedition. His work continued to fascinate the king, the court (especially the faithful Colbert), and the public. He became, for a while, *the* dominant figure in astronomy in Europe. Like Galileo, Cassini was astute when it came to self-promotion. More than anyone else in his time, he used the publication of papers in scientific periodicals as a way of establishing priority and of letting the educated world hear of his successes.

In 1676, Jupiter's moons made possible another measurement that would be essential to later astronomy. Ole Roemer, a Danish astronomer also working at the Royal Observatory in Paris, studied the eclipses of Jupiter's moons and noticed that the time that elapsed between their disappearances behind Jupiter varied with the distance between Jupiter and the Earth, a distance that changes as the two planets move in their orbits. Roemer speculated that the velocity of light was responsible for what looked like delays in the eclipses. When Jupiter was farther from the Earth, it took longer for the picture of the eclipse to reach eyes and telescopes on the Earth. Timing the delays, Roemer proceeded to calculate the speed of light at about 140,000 miles per second. That figure fell short of the 186,282 miles per second that is assigned to the speed of light in a vacuum today. The discrepancy in the measurement owed in part to Roemer's less-than-precise knowledge of the distance to Jupiter.

The modern calculation is made with an atomic clock and a laser beam. (This book uses rounded-off figures of 186,000 miles or 300,000 kilometers per second for the speed of light.)

Louis XIV was not the only monarch who patronized astronomy in the late seventeenth century. The impetus for the founding of the Royal Observatory in England came, however, from an unlikely source. In 1674, Louise de Kéroualle, a Bretonne who was one of King Charles II's mistresses and had recently been made duchess of Portsmouth, brought a Frenchman named Sieur de St. Pierre to the king's attention. St. Pierre had discovered, so he said, a secret method for finding longitudes. Determining longitude requires having some form of universal clock—such as Jupiter's moons were for Cassini and Richer—that will allow comparison of celestial phenomena as seen from different locations. If the Sun is directly overhead in New York, what is its position in Greenwich, England? Answering such questions is one way of finding longitude.

Flamsteed and others thought that St. Pierre's "secret method" must involve using the Moon as a timekeeper. Who needed St. Pierre? King Charles told the Royal Society, then known as the Royal Society of London for Improving Natural Knowledge, to collect whatever lunar data were necessary to determine whether the Moon could serve as a universal clock. Flamsteed soon reported that lunar and stellar positions were not known sufficiently well to make the method reliable. Nevertheless, the bee was in the king's bonnet, and he founded the Royal Observatory, designating Flamsteed as "Astronomical Observator," a position that would later become "Astronomer Royal." A new building, designed by Sir Christopher Wren, was ready for occupancy in July 1676, and Flamsteed and his associates began the task of producing correct star positions and tables for the Sun, Moon, and planets—for the express purpose, as those who governed England saw it, of aiding navigation, but also for the furthering of astronomy.

So things stood in the last quarter of the seventeenth century, with kings footing the bill for observatories in Paris and at Greenwich and sponsoring such bodies as the French Academy of Sciences and the Royal Society, where gentlemen—expert and not—got together and shared knowledge of the wonders of science. These gentlemen had a fairly good idea of the dimensions of the solar system. Astronomy seemed poised and ready to measure much greater distances. But a baseline from Paris to Cayenne wasn't long enough to measure the parallax shift of stars; in fact, no possible baseline on the face of the Earth would suffice. Was there any way to make the parallax method work for stars?

The clue was there long before Ptolemy. One objection to Aristarchus's suggestion that the Earth moves and orbits the Sun was that, if it does, observers on Earth ought to be able to observe stellar parallax as the Earth travels from one extreme of its orbit to the other. In other words, the stars ought to shift position in the sky. If you are viewing the cactus and the mountains from a car window and they don't seem to be changing positions relative to one another, either your car is standing still or both the cactus and the mountains are *very* far away. Likewise, if there is no observable stellar parallax, either the Earth does not move and orbit the Sun, or the stars are extremely distant—more distant than most ancient scholars were willing to conceive. Aristarchus of course had answered the objection by saying that the stars must indeed be infinitely far away.

By 1700, nearly all astronomers agreed that the Earth rotates on its axis and orbits the Sun. They also recognized the potential of this orbit for a baseline. Even so, at that time and even much later, no one was able to detect an annual shift in a star's position. That discovery had to wait for considerable improvement in telescopes and the precision instruments used with them—advances that would eventually come with the Industrial Revolution. Meanwhile, there was also more to be learned from a theoretical point of view, in order to be able to recognize stellar parallax and not confuse it with other effects.

Deciphering Distant Light

According to the inverse square law, the brightness of a light falls off with distance in a mathematically dependable way. The measured intensity of light diminishes by the square of the distance to its source. Imagine that you have two 100-watt lightbulbs. Place one of them twice as far away as the other. The farther bulb will appear to be only a fourth as bright as the nearer. It will seem to you that it must be a 25-watt bulb. Turning that exercise around: Suppose you keep one of the 100-watt bulbs nearby and ask a friend to carry the second to an unknown distance. If the more distant bulb appears to be a 25-watt bulb, you can be sure it's twice as far away as the first bulb. Comparing their *apparent* brightness gives the distance of the second light. How does this apply to stars? If stars all had the same close-up brightness, and the distance to one star were known, then in principle it would be possible to find the distance to any other star by comparing its *apparent* brightness (how it looks from the Earth) with the apparent brightness of the first star.

The same inverse square law that works with the brightness of light works with the size of objects. Even without knowing the mathematics, you and I use the method instinctively, if imprecisely. Suppose you are overlooking a vast expanse of land with a great number of elephants grazing on it. Some of the elephants look tiny, but because you have previous experience of elephants and know how large an elephant would be if you were standing near it, you make an educated guess that these aren't really tiny at all, just far away. From the difference between their apparent sizes and the normal close-up size of an elephant, you judge how far away they are.

Your success relies upon (1) having some way of knowing that elephants are all approximately the same size—some reason for assuming there aren't elephants that are pygmies and others that are giants—and (2) having experience of what that size actually is, that is, how big an elephant looks when it's standing a known distance away.

With a light seen at night at an unknown distance, on the Earth or in the sky, those essential ingredients are lacking. Experience teaches that it's not possible to depend on all lights having approximately the same close-up brightness, so knowing the close-up brightness of one light is of no use when studying the brightness of a distant light. Comparison is meaningless. Is it a dim bicycle headlamp over there, just a few yards up the road, or is it the high-beam headlight of a car a mile away? Is it a meteor suffering a fiery death in the upper atmosphere, or is it a firefly no farther away than the treetops?

The situation is puzzling but not completely hopeless, nor did it seem so at the end of the seventeenth century. Astronomers did know how far it is to one star—the Sun. All stars including the Sun *might* be equally bright. It would seem highly desirable to know the brightness or the distance of at least one star other than the Sun, but failing that, the Sun would have to serve as a standard. On the assumption that it could, Englishman Isaac Newton set out to measure the distances to the nearest stars.

Newton was born the year that Galileo died, 1642, and was thirty years old when Cassini and Flamsteed measured the distance to Mars. Except for his *Principia Mathematica*, which appeared in 1687 and is unarguably one of the most important achievements in the history of Western thought, Newton published almost nothing. However, he obsessively researched an astounding variety of subjects, including optics, theology, alchemy, and calculus, and was responsible for significant advances in many of the fields he investigated. He contributed to the development of a new type of telescope that used a mirror rather than a lens to focus incoming light. (See figure 4.5, page 141.)

Newton probably wouldn't even have published the *Principia* had his friend Edmond Halley not goaded him sufficiently. Publication meant public notice, and contact and correspondence with other scholars. He would be expected to take part in the discussions and arguments at the Royal Society. There would be invitations to perform experiments for that body and to watch others do the same. Though honors were tempting and sometimes Newton succumbed, those distractions more often seemed anathema to him, for they took precious time away from his research. It is surprising, therefore, that he did agree to become head of the Royal Society. Unfortunately, he used the position of power in an extremely unpleasant, autocratic manner, bringing grief to other fine scientists (including the elderly Flamsteed). Newton also eventually accepted the job of "Master of the Mint" and thoroughly enjoyed his rather pedestrian duties there.

Whatever his personal failings, it was Newton who capped off the Copernican revolution with his discovery of laws of gravity. In Newton's description, each body in the universe is attracted toward every other body by the force called gravity. How much bodies are influenced by one another's gravitational attraction depends on how massive the bodies are and how near they are to one another. For instance, any change in the mass of the Earth or the Moon, or in their distance from one another, would change the strength of the gravitational attraction between them. If the mass of the Earth were doubled, the attraction between the Earth and the Moon would double. If the Moon were twice as far from the Earth as it is, the attraction of gravity between the Earth and the Moon would be only one-fourth as strong.

Here at last in this simple description were the dynamics that cause the planets to move as they do rather than in some other way—the physical reasons behind Kepler's laws. The answer that eluded Ptolemy, Copernicus, Galileo, and Kepler—but to which Kepler's laws point—was summed up in one sentence: The gravitational force between any two bodies is proportional to the product of their masses and inversely proportional to the square of the distance between them. It took Newton's genius to see that it is this same force—gravity—that keeps us from flying off the ground, dictates the path of a ball thrown on the Earth, underlies the motions of the planets, and governed the way Galileo's two objects landed at the foot of the Leaning Tower of Pisa (if they did).

Newton's "laws of motion" could be tested by experiments as well as by astronomical observations, and this was an era that rejoiced in such testing. The scientific rigor that had made Galileo exceptional was beginning to be considered essential for any scientific activity. It became clear that Newton's formulas did indeed describe the way things happen. Nature was law abiding, and it was following these laws! It's difficult for us, who take for granted that science is able to predict reality and that simple, dependable mathematical and scientific rules underlie the apparent complications of nature, to appreciate how awe inspiring it was for the many men and women who were realizing this in the seventeenth century for the first time. Not only did such laws exist, but human minds could discover them and understand them. The concept was not a complete novelty to scientists, although to see it demonstrated so beautifully *was* a novelty. To the nonscientific public, Newton's revelation was sensational, miraculous. The fame of *Principia* spread quickly throughout Europe, and Newton's ideas were popularized in many forms. A publication called "Newton for Ladies" appeared in France. Newton's *Principia* did encounter some hostility on philosophical grounds, uneasiness with the notion that gravitation could act through empty space. "Action at a distance" suggested the occult.

Newton's attempt to estimate the distances to the nearest stars by assuming that all stars, including the Sun, have approximately the same brightness is one of his less celebrated efforts. In the analogy with the elephants, you assumed that all elephants are about the same size, but if you had seen only one elephant close up, it would have been risky to jump to the conclusion that your local elephant was a typical elephant. The difference in apparent size between it and other elephants whose distance you didn't know might actually have been due to differences in size, not an indication of their distance at all. Newton's situation was even more ambiguous than that. Elephants look pretty much alike, but the Sun, as seen from Earth, doesn't resemble the other stars. Most experts in Newton's time did think that the Sun was a star, but was it a *typical* star? It wasn't unreasonable to decide to assume that it was, and see where that would lead.

Newton proceeded in an ingenious but rather convoluted way, using a technique suggested by Scottish mathematician and astronomer James Gregory. Judging by the size of Saturn, Newton estimated that about one part in a billion of the Sun's light hits the planet Saturn. He reasoned that Saturn doesn't reflect all of the sunlight that hits it, so it would be incorrect to conclude that the light seen coming from Saturn represents one billionth of the Sun's light. Instead, Newton thought Saturn probably reflects only about a quarter of the Sun's light that hits it, which would mean that the reflected sunlight coming from Saturn gives a good indication of what one part in 4 billion of the Sun's light looks like. Following this line of thought, if a distant star seems to have the same brightness as Saturn, it follows that the light coming from that star (not reflected sunlight; the star's own light) is also equivalent to one part in 4 billion of the Sun's light. What this means—*if* all stars' brightnesses are the same as the Sun's—is that a star that looks (from Earth) as bright as Saturn would have to be approximately 100,000 times farther away from Earth than the Sun is.

Newton's measurements of the distances to some of the nearest stars were not far wide of the mark, though this method was undependable partly because stars do differ greatly in their absolute mag-

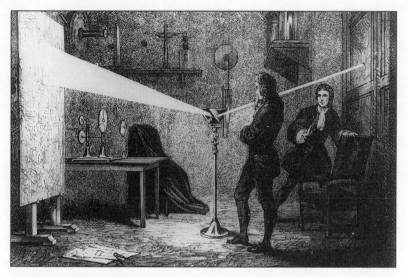

Woodcut of Newton demonstrating his prism experiment. *(Corbis)*

nitude (their "close-up" brightness), and a star's apparent magnitude (how bright it appears from Earth) alone can't be used as a gauge of its distance. (See the box.) Some of the stars that look brightest in the night sky are very far away, while many nearer stars are rather inconspicuous.

The brightness with which a star would appear if it were only ten parsecs distant from us—what we might call its "close-up brightness"—is its

absolute magnitude

How bright a star appears from the Earth as we view it in the night sky is its

apparent magnitude

It is the apparent magnitude of a star that falls off with distance. The absolute magnitude of a star does not change with distance.

A **parsec** is a little more than 30 trillion kilometers or 3.26 light-years.

A **light-year** is the distance that light travels in one year—about 5.88 trillion miles or 9.4607 trillion kilometers.

In 1718, Newton's younger friend Edmond Halley discovered an important new clue to star distances. He'd become fascinated with Ptolemy's writings and star catalogs. Halley was particularly curious about whether the stars had changed positions since the time of Hipparchus and Ptolemy, and he took it upon himself to compare positions he observed in the late seventeenth and early eighteenth centuries with positions recorded in the *Almagest*.

Halley was born in 1656 and while still an undergraduate at Oxford wrote and published a book on Kepler's laws. His book came to the attention of Flamsteed, who had a great deal of influence as the new Astronomical Observator and first head of the Royal Observatory. Halley left Oxford without getting his degree and, at Flamsteed's behest, was soon on the island of St. Helena in the South Atlantic, mapping the sky as seen from the Southern Hemisphere. When he returned to England two years later, he was elected to the Royal Society. He was only twenty-two.

Even more eclectic in his interests than Newton, Halley spent the next thirty years in an astounding variety of pursuits. He traveled extensively to meet other scientists and astronomers; he assisted Flamsteed; he got married; he commanded a warship in the Royal Navy and captained a mutinous ship across the Atlantic; he went to Vienna on two secret diplomatic missions; he served as deputy to the controller of the Mint at Chester (a position Newton helped secure for him); he studied magnetism and the winds and tides; he prevailed upon Newton to publish the *Principia*, and he financed its publication. It was Halley's study of comets that brought him most fame. (After his death, Halley's comet was named for him when it reappeared in 1758 at the time he had predicted.) In 1703, Halley joined the faculty of the University of Oxford, where he had failed to complete his degree, as Galileo had done at the University of Pisa.

In 1718, Halley reported that three of the stars he was studying—Sirius, Arcturus, and Aldebaran—had shifted over the centu-

Engraving of the Octagon Room of the Royal Observatory in Greenwich, England, as it looked in Flamsteed's time. *(probably by Francis Place)*

ries since Ptolemy's time. Halley strongly suspected that the discrepancies between the old charts and those of his own era were too large and too isolated to be attributed to errors in ancient measurement. Why should early astronomers have got everything else right but this? His suspicions were confirmed when he measured the shift of Sirius during the one hundred years since Tycho Brahe had observed it. The change had been so gradual that it could only be noticed over a span of at least several human generations.

"Proper motion" is the name given this movement of stars relative to one another over the centuries—an apparent movement across the sky when viewed from the Earth. Most stars are not moving only side-to-side, of course, as though the sky were a two-dimensional surface; they are likely at the same time to be getting closer to Earth or farther away.

In 1720, at age sixty-four, Halley succeeded Flamsteed as Astronomer Royal, which would not have pleased Flamsteed, for Halley had acquiesced in Newton's poor treatment of the old man. It probably *would* have pleased Flamsteed that his widow whisked all the instruments out of the Royal Observatory. They were legally hers because the financial arrangement at the Observatory was such that the Astronomer Royal purchased equipment out of his salary. Halley had to acquire new instruments. He died in 1742 at the age of eighty-five. One of his most significant achievements didn't come to fruition until nearly twenty years later.

Halley had put little faith in earlier measurements of the Sun's and the planets' parallaxes and distances, including those of Cassini and Flamsteed, although his friend Newton came to agree with them. Newton also measured the orbits and the distances of the planets from the Sun, using the dynamics of the system rather than astronomical observations as the basis for his calculations. He concluded that Cassini's and Flamsteed's results were better than his own. But as of 1700, the only real consensus among astronomers when it came to the Sun's distance was that the Sun was at least 55 million miles away.

Now when Halley had been on St. Helena in his early twenties, he'd seen and timed a transit of Mercury across the Sun. Halley realized that a transit would provide an opportunity to use parallax in a new way to measure distances in the solar system. The passage of a planet across the Sun shows up as a tiny black dot passing across the Sun's face, and observers at different locations on the Earth's surface see the planet first touch the Sun's disk at different times. A transit is a relatively rare event, but Halley knew there would be a transit of Venus in 1761. He also knew he wouldn't be alive to witness it unless he lived to be 105. So he wrote and published detailed instructions on the best way to use observations of the transit from different parts of the world.

Sixteen years after Halley's death, with the return of the comet he'd seen in 1682, his name became a household word. In 1761 there was indeed considerable effort, much of it due to his prestige, to study the transit of Venus, and similar excitement about a second transit in 1769. Observers, in many locations around the world, knew the opportunity wouldn't be repeated again until 1874. There are colorful stories connected with this venture, which indicate that many astronomers at the time were less denizens of the ivory tower cum telescope than they were prototypes of Indiana Jones.

Frenchman Guillaume le Gentil planned to observe the 1761 transit from Pondicherry, near Madras in India. He arrived to find the town occupied by British forces. This was during the Seven Years' War, England and France were enemies, and le Gentil was not welcome in Pondicherry. Rather than turn around and head for home, he settled nearby for eight years, supporting himself in part by trading while he waited for the next transit. By then the British had ceased to be an obstacle, but nature had no mercy on le Gentil. The Sun shone brightly before and after the transit. During the transit, it was hidden by a cloud.

Jean d'Auteroche led another French observing team in Russia in 1761, and in what is now southern California in 1769. The party of four astronomers trekked overland across Mexico to reach their California observing location. D'Auteroche and two of the other men died of disease shortly after their arrival. That left the fourth to undertake the treacherous return journey alone, but he brought back with him the dearly bought records of the observation.

The Reverend Nevil Maskelyne, sent by the Royal Society to St. Helena to observe the 1761 transit, had a much better time of it. Maskelyne's expenditures were in the neighborhood of £292, out of which £141 went for his personal liquor supply.

David Rittenhouse, in America, labored for months before the 1769 transit building a temporary log observatory at Norriton, near Philadelphia. He used a collection of instruments there that included telescopes from Europe and others he had built himself, and also his own eight-day clock that "does not stop when wound up, beats dead seconds, and is kept in motion by a weight of five pounds."

Charles Mason and Jeremiah Dixon, who would later establish the Mason-Dixon line, headed up another team sponsored by the Royal Society. That august body threatened them with disgrace and possible legal action if they failed to continue with their expedition to the Cape of Good Hope to observe the 1761 transit, after a French frigate attacked their ship in the English Channel and eleven crew members died. Evidently the captain of the French frigate had been unaware that in spite of the ongoing Seven Years' War, Englishmen and Frenchmen were collaborating in this scientific endeavor.

Maximilian Hell, a Viennese Jesuit astronomer, observed the 1769 transit from Norway and suffered devastating damage to his reputation when Jerome Lalande insinuated that Hell had fiddled with his observations to make them consistent with those reported by others. Karl von Littrow supported Lalande's allegation, claiming to have found proof in the form of different ink colors in Hell's report. Hell's good name wasn't restored until 1883. Among other things, it was discovered that von Littrow had been color-blind.

Sadly, the results of all this effort were less definitive than Halley and these astronomers had hoped. Precise determination of the instant the planet touched the Sun's disk was much more difficult than anticipated. Because of the Sun's corona and the atmosphere of Venus, Venus's image at the beginning and end of the transit was blurry. Calculations based on the results of these observations put the distance from the Earth to the Sun at about 95 million miles or 153 million kilometers, as compared with Cassini's measurement in 1672 of 87 million miles or 140 million kilometers, and the modern measurement of 93 million miles or 149.5 million kilometers.

Halley's discovery of proper motion was also destined to bear fruit far beyond his lifetime. It so happens that the three stars whose proper motion he first measured—Sirius, Arcturus, and Aldebaran—are some of the brightest in the sky. Was this mere coincidence? A star might look brighter than others because it really is brighter, or it might look brighter because it is closer. Halley's discovery of proper motion gave researchers a new clue.

Objects moving across our line of vision close to us appear to move more rapidly against the background than those farther away. A child on a tricycle nearby can easily outrace an automobile driving at a good clip off on the horizon. Logic says that the same will be true with stars that are moving across an observer's line of vision. Unless stars are all the same distance from Earth, nearer stars should appear to move against a background of more distant stars. Most stars, in fact the vast majority of them, have shown no change of position since Ptolemy's time for a viewer on Earth. Does that mean they are far more distant than those that have changed position?

Sixty-six years after Halley's discovery, astronomer William Herschel, the discoverer of the planet Uranus, studied the proper motions of a number of stars and the way those proper motions relate to one another, and from that he was able to plot the Sun's motion through our part of the Galaxy. Still later, the German astronomer Friedrich Wilhelm Bessel, suspecting that proper motion, rather than brightness, might be the most significant indicator of which stars are nearest, used it as a basis for choosing which stars to try to measure with the parallax method.

Annual stellar parallax means the apparent shift in a star's position caused by the observer's traveling from one extreme to the other of the Earth's orbit. As Earth returns in its orbit to where the observation began, the star also returns to its original position.

Proper motion is the change in a star's position over a much longer period of time, due to the fact that all stars including the Sun move within the Galaxy. Proper motion was discovered more than a hundred years before anyone was able to detect annual stellar parallax.

The stellar parallax shift that ancient Greek and Hellenistic astronomers could not find does exist, just as astronomers around 1700 were sure it must. But the shift is extremely tiny and difficult to detect. Certainly it isn't possible to see it with the naked eye, so the astronomers in antiquity can't be blamed for missing it. Telescopes of Galileo's and Cassini's time weren't refined enough to reveal it either.

One man who attempted to detect stellar parallax—making other important discoveries in the process—was James Bradley, born in England in 1693. The star Gamma Draconis passes almost directly overhead in London, and Bradley and his friend Samuel Molyneux, a wealthy amateur astronomer, decided to try to measure its parallax motion. They attached a twenty-four-foot-long telescope to a stack of brick chimneys on the building where Molyneux was living. By using a screw, they could adjust the telescope to keep it tilted toward the star. The result was puzzling. Instead of having to adjust the tilt most in December and June, as they had expected, the adjustment was most extreme in March and September and was so large that, even if it had occurred at the right time of year, it was highly unlikely to be caused by parallax. Bradley took advantage of an exceptionally understanding aunt, who allowed him to cut holes in her roof and floors and install a larger and more sophisticated telescope. Observations with this instrument only repeated his and Molyneux's earlier baffling findings.

The explanation dawned on Bradley, so the story goes, while he was taking a cruise on the Thames River. When the boat changed direction, a weather vane on the mast shifted. It wasn't the wind direction that had changed, however. It was the boat's direction in relation to the direction from which the wind was blowing. Bradley realized that the displacement of the stars he was studying was similarly caused by the changing motion of the Earth. Just as the wind direction seemed to shift according to the direction the boat was moving, so starlight seemed to shift according to the direction of the Earth's motion.

What Bradley's telescopic observations demonstrated was that the Earth orbits the Sun and the speed of light is not infinite, both of which were already well accepted. Bradley gave the name "aberration" to the effect he had found and announced the discovery to the Royal Society in 1729. Aberration produces a 20^1/$_2$ arcsecond shift (see figure 4.4) in the apparent positions of stars over a year's time. Bradley also found that the Earth wobbles due to the fact that its shape is not perfectly spherical, and he gave the wobble the name "nutation." Aberration and nutation were not what Bradley had set out to find; but these discoveries were actually helpful steps on the road to discovering the tiny displacement of true annual stellar parallax. In any search for annual stellar parallax, one needed to take into account these other reasons why the positions of stars change with the seasons. The negative side of Bradley's discoveries was even more significant. He had shown that the parallaxes of stars could not be more than one second of arc, for had they been as large as one arcsecond, he knew he would have been able to detect them. This meant stars were much farther away than was generally supposed.

In 1742, Bradley succeeded Halley as Astronomer Royal.

The Triumph of Celestial Mechanics

The eighteenth century saw a rapid increase in the number of observatories in Europe. Among the sciences, only in medicine were there more people professionally involved in research. Not all of them were peering through telescopes, for the meticulous cataloging of what others found required many clerical workers. Funding came from governments, universities, scientific bodies, even religious soci-

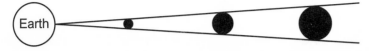

Figure 4.4 The arcsecond originated as an ancient Mesopotamian measurement. A circle has 360 degrees. Each degree can be divided into 60 minutes of arc or "arcminutes." Each minute of arc can be divided into 60 seconds of arc or "arcseconds."

The measurement of an arcsecond isn't a measurement of true size. If you hold your finger up at arm's length against the sky, its width covers about two degrees of arc. But this finger held at arm's length would cover a branch on a nearby tree, the Concorde flying overhead, the entire Moon, or (out of sight with the naked eye) an enormous number of galaxies. Clearly not all objects in the sky that have the "angular size" of two degrees of arc are the same true size. How "large" two degrees of arc are depends upon how deep into space you're looking. The Moon's angular size is about half a degree of arc. The Sun's angular size is approximately the same as the Moon. Yet these two bodies are definitely not the same true size.

In the picture above, each of the circles has the same angular size viewed from Earth, covering the same number of degrees of arc, yet they are not the same true size. (Recall Aristarchus's study of the Sun and Moon, shown in Figure 1.4.)

One arcsecond is the angular size the width of your finger would have if you were able to hold your finger up about 5,000 feet (1,500 meters) above you.

eties—the expense often justified in terms of the practical spin-off for navigation, mapping, and surveying, and the prestige such institutions gave to their sponsors.

The educated public loved astronomy and followed it avidly. Traveling lecturers drew large audiences; less technical books were popular and telescopes for amateur use sold well, as did globes. Natural theology—the argument that nature and the harmony of the universe are eloquent proof of the existence of God and indicate what God is like—was preached from many pulpits, given voice in hymns and secular poetry, and favorably discussed in academic circles. Though astronomy had ceased to be a required part of the curriculum in most universities, it was widely considered essential to a proper, gentlemanly education.

In England, the Royal Observatory continued to be financed from government coffers, a considerable investment. The Observatory's demand for equipment, as well as the Royal Society's great interest in the improvement of all scientific instruments including telescopes, helped support and encourage a healthy local optical industry. The Industrial Revolution, which started in England in the eighteenth century, brought advances in the design and construction of machines and in precision engineering in general. Astronomy reaped enormous benefits from this progress and also, with the increasing expertise of its instrument makers and demand for their products, contributed substantially to it. The finest telescopic instruments came from English manufacturers, who were suppliers to all Europe, setting the standard and style of the profession. Earlier astronomers had often designed and built their own instruments, and some of the more notable among them would continue to do so, but increasingly there was a division between those who manufactured telescopes and those who used them. Telescope builders were not necessarily considered a lower breed than practicing astronomers. Some were elected to the Royal Society.

In the early nineteenth century, the Industrial Revolution spread to Europe and the United States. London opticians were still producing reflecting telescopes, the type pioneered by Newton and others in the late seventeenth century (see figure 4.5), at affordable prices, and English amateurs were making some excellent instruments themselves. However, Britain's prime minister William Pitt and his government dealt London's optical industry an almost mortal blow by imposing a punitive tax first on windows and a little later on all glass. After Swiss glassmaker Pierre Guinand moved from Switzerland to Munich in 1804, bringing with him a new method of mixing molten glass, Germany took the lead in refining the art of telescope design and manufacture. Bavarian-born Josef von Fraunhofer, who had worked for a time as Guinand's assistant, greatly improved the refracting telescope (again, see figure 4.5). Friedrich Wilhelm Bessel pioneered the use of the meridian circle, which made it possible to measure two coordinates of a star at the same time, adding significantly to the accuracy of observations, as

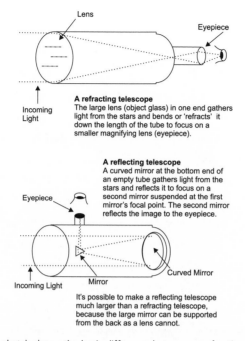

Figure 4.5 This sketch shows the basic difference between a refracting and a reflecting telescope. There are many varieties of each.

did advances in the design of astronomical clocks. Eighteen-year-old mathematician Karl Friedrich Gauss, at Göttingen in 1804, invented the method of "least squares," which enabled an astronomer to choose the best observations in a less arbitrary manner.

Finally, by the late 1830s, improved technology and better theoretical understanding had converged to the extent that it was possible for three astronomers to detect annual stellar parallax. Clearly, it was a discovery whose time had come. The three measurements occurred independently but almost simultaneously.

Bessel, in Königsburg, Germany, was the first to announce his findings, in 1838. Reasoning that proper motion, rather than brightness, might be the most significant indicator of which stars are nearest, he chose 61 Cygni, a dim star with a large proper motion (5.2

arcseconds a year; see figure 4.4). The figure at which he arrived for its annual parallax was 0.3136 arcseconds. Knowing the distance the Earth had traveled in its orbit to produce this displacement allowed him to calculate the distance from the Earth to the star—3.4 parsecs (11.2 light-years)—which is, for comparison, 600,000 times greater than the distance from the Earth to the Sun. Bessel's measurement was a considerable step up on the cosmic distance ladder.

Meanwhile Thomas Henderson from Scotland, observing from South Africa, decided to study Alpha Centauri, choosing it on the basis of brightness rather than proper motion. Alpha Centauri is the third brightest star in the night sky. Though Henderson actually measured Alpha Centauri's parallax before Bessel measured 61 Cygni's, Henderson didn't announce his results until he got back to Britain early in 1839—thus losing out to Bessel in the pages of history. A year later, Friedrich von Struve, born in Germany but working in Tallinn (now in Estonia), announced that he had measured the parallax for Alpha Lyrae (aka Vega), the fifth brightest star in the night sky.

The parallaxes measured for all three of these stars were small:

* For 61 Cygni, a parallax of 0.3136 arcseconds (see figure 4.4), a distance from the Earth of 3.4 parsecs (11.2 light-years). Astronomers now know that 61 Cygni is a double star.
* For Alpha Centauri, about 1 arcsecond parallax—a figure later refined to 0.76. The Alpha Centauri *star system* (for it consists of three stars) is 1.3 parsecs away (4.3 light-years). The star Proxima Centauri is one of that trio and at the moment is the solar system's closest neighbor. It revolves around its mates Centauri A and B every 500,000 years.
* For Alpha Lyrae (Vega), a parallax of 0.2613 arcseconds, a distance of 8.3 parsecs (26 light-years).

As far away as they are, these stars are some of the nearest to Earth. With their measurement, it began to sink in how profoundly alone our little solar system community is. After Cassini and

Flamsteed, the solar system had seemed so huge. Now it shrank almost to the vanishing point in comparison with the enormous emptiness a space traveler would have to cross to reach anything else beyond. You must multiply the distance from the Sun to Pluto (the planet farthest from the Sun) by 9,000 to reach Proxima Centauri—the next break in the darkness.

The successful measurement of the distance to the nearest stars strengthened an impression that had existed since the eighteenth century that "celestial mechanics," the marriage of mathematics and astronomy, was the highest of all the sciences and the most valuable for deepening human understanding of the laws of nature. Another achievement crowned this reputation even more spectacularly. Soon after Bessel, Henderson, and Struve's measurements, Urbain-Jean-Joseph Leverrier, who was the virtual dictator of the Royal Observatory in Paris and scorned and discouraged any intellectual pursuit that *wasn't* celestial mechanics, studied the orbit of the planet Uranus. He came to the conclusion that certain mysterious irregularities in the orbit that do not accord with Newton's laws must be caused by the gravitational pull of another undiscovered planet. On September 23, 1846, Johann Gottfried Galle at the Berlin Observatory, looking for that unknown planet where Leverrier had predicted it should be, and working with records kept by British astronomer John Couch Adams, discovered Neptune. Leverrier's prediction had been astoundingly close to right—quite by coincidence, actually, for he had not chosen correctly among several possible solutions for the orbit. The discovery was a public sensation. A miracle! Indeed, more of a miracle than the public knew, given Leverrier's wrong choice.

Reading Between the Lines

At midcentury, astronomy and celestial mechanics did indeed seem to be moving from triumph to triumph. They had also suffered one setback, for knowing the distances to a few stars gave astronomers a way to calculate their absolute magnitudes and erased forever any hope that the absolute magnitude of all stars is the same. If you were to find that the elephants on the plain come in a variety of sizes, and you could measure directly the distance to only a few of them, how would it be possible to judge the size or distance of the others? One approach would be to take another look at this category you've been calling "elephant" and see whether it can be broken down. Maybe there are Indian elephants and African elephants, and some way to tell them apart. If all elephants aren't the same size, perhaps all Indian elephants *are*.

The trick would be to find characteristics that (unlike apparent size) won't change with distance, such as distinctive ears. You could call those with extraordinarily large, wavy ears Group A elephants. Suppose you do have a way of measuring the actual distance to a few of the Group A elephants and discover that their size doesn't vary greatly. It seems fairly safe to assume that those Group A elephants too far away for direct measurement are also that size. With that assumption, take the exercise farther. If another animal is standing near a Group A elephant in the distance, drinking from the same water hole, you can judge the size of that second animal by comparing it to the elephant. Suppose the second animal is spotted and has an extraordinarily long neck. Name it a giraffe. Now if you go on watching elephants and giraffes out there and conclude that all giraffes are about the same size, you have a way to calculate the distance to any other animal that appears to be a giraffe. One measurement builds on another. Of course, if there is a mistake somewhere along the line—maybe Group A elephants come in two sizes, or maybe the animal you think is near the elephant is actually fifty feet beyond it—the whole measurement structure begins to collapse and has to be recalibrated.

Stars don't all have the same absolute magnitude, but the hope was that those within certain recognizable categories *do* and that these categories would be recognizable by other distinguishing characteristics that don't change with distance. Even before the first stellar parallax measurements, there had been some developments that would lead to a better understanding of stars.

At the beginning of the nineteenth century, it had been generally assumed that it would never be possible to discover the chemical composition of stars or their physical makeup, because researchers couldn't get near enough to examine them. The French philosopher Auguste Comte pointed to the chemical composition of stars as an example of "unobtainable knowledge." Not everyone shared this pessimism. Researchers were soon to find that starlight carries with it an enormous amount of information about its source, if you can crack the code.

Since Newton's study of optics, both scientists and laypeople had known how to use a glass prism to break a ray of light into its component parts. When white light passes through the prism, the colors of which the light is composed spread out in an ordered sequence—the *spectrum*—the familiar rainbow. The order is always the same: red, orange, yellow, green, blue, indigo, and violet. The acronym for that is "Roy G. Biv."

Position in the spectrum is referred to by colors ("the red end of the spectrum" or "the violet end of the spectrum"), or, more precisely, by wavelengths, because each color is produced by a different wavelength of light. The longer waves are the red. The waves grow shorter as you move across the spectrum to violet. (See figure 4.6.)

Light that human eyes can see—the visible spectrum—is only a small part of the much larger electromagnetic spectrum. What is out beyond red on the one hand and violet on the other is invisible to us, but there is a great deal out there: Infrared light, microwaves, radio waves, ultraviolet light, X rays, gamma rays—all of them are

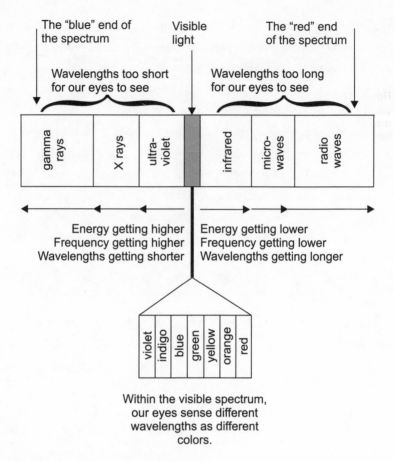

Figure 4.6 The Electromagnetic Spectrum

forms of electromagnetic radiation, with wavelengths either too short or too long to be within the visible spectrum.

When light passes through a prism, the resulting spectrum provides information about the light source, even when that source is billions of light years away.

- An incandescent, solid light source radiates all colors, and the spectrum is continuous from violet to red (and beyond the visible spectrum in either direction).

Figure 4.7 Absorption Spectra. When an incandescent solid is surrounded by a cooler gas, the result is a spectrum in which a continuous background is interrupted by dark spaces—called "absorption lines"—that occur because the gas has absorbed from the light those colors which the gas would radiate itself.

- An incandescent gas radiates only a few isolated colors, and each kind of gas has its own pattern, called an emission spectrum.

- When an incandescent solid (or its equivalent) is surrounded by a cooler gas, the result is a spectrum in which a continuous background (such as the spectrum an incandescent solid would produce) is interrupted by dark spaces—called absorption lines. (See figure 4.7.) In this case the gas surrounding the original light source has absorbed from that light those colors that the gas would radiate itself, and this kind of spectrum is called an absorption spectrum. By looking at the pattern of the absorption lines and noting where they fall within the spectrum, one is able to discern which gas or gases are responsible for the absorption.

Much of the modern understanding of light and spectra stems from the pioneering work of Josef von Fraunhofer. Born in Straubing, Bavaria, in 1787, Fraunhofer was the eleventh and youngest child of a master glazier and worker in decorative glass. Orphaned at twelve, he became the apprentice of a mirror maker and glass cutter in Munich, who paid him nothing, offered minimal instruction, and made it impossible for him to attend the Sunday Holiday School, which offered apprentices a little schooling outside their trade. Fraunhofer's luck turned for the better when he was fourteen and his master's house collapsed, burying the boy in the ruins. His escape—he was injured but protected from death by a crossbeam—became a news item in Munich and reached the sympathetic ears of the elector Maximilian. Maximilian gave young Fraunhofer some money, which he used wisely, buying out of his apprenticeship and purchasing a little equipment for himself. When his own business (engraving visiting cards) failed to support him, he had to go back to work for his old master, but only temporarily. Fraunhofer's miraculous survival also drew the attention of a wealthy Munich lawyer and financier named Utzschneider, who soon hired Fraunhofer to work at his glassmaking establishment. Such was Fraunhofer's innate ability and zeal for his craft that when he was in his early twenties Utzschneider had already put him in sole charge of the glassworks.

Fraunhofer was one of a handful of men in the early nineteenth century who rose from working-class backgrounds to become leaders in astronomy. In a short lifetime, he designed and built increasingly fine telescopes, among the best in the world at that time, and he was responsible for a number of inventions that made their use more effective. Bessel and Struve were using Fraunhofer telescopes when they first measured stellar parallax. Fraunhofer's discoveries about light led to some of the most significant developments of the nineteenth and twentieth centuries, making him one of the most important figures in the history of optics. He was the first to study and map the absorption lines of the Sun's spectrum.

In 1814, Fraunhofer was trying to find ways to make more accurate measurements of the way a piece of glass refracts the light. When Newton had studied the spectrum of light, he'd done so by allowing sunlight entering a round hole in a shutter to pass through a glass prism and fall on a screen. Fraunhofer used a modification of Newton's experiment. Basically, what he did was to substitute a narrow slit for the round hole and a telescope for the screen. Fraunhofer found that the continuous spectrum of the Sun is interrupted by many dark lines, and he found the lines

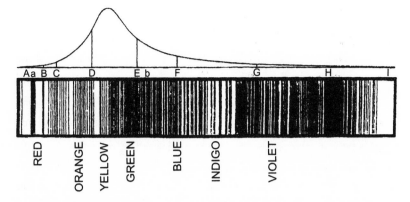

Figure 4.8 Fraunhofer's map of the solar spectrum. He used the letters to identify the most prominent absorption lines.

present in *all* sunlight, whether direct or reflected from other objects on Earth or from the Moon and planets. He labeled the 10 strongest lines in the solar spectrum and recorded 574 fainter lines. (See figure 4.8.)

Continuing to investigate, Fraunhofer found dark lines also appearing in the spectra of stars, but in different arrangements. He concluded that the lines must originate in the very nature of the Sun and the stars. What he had actually discovered were the signatures of the different chemical elements present in the Sun's and stars' atmospheres, the third type of spectrum described above—an absorption spectrum (the type in which the cooler gas surrounding the original light source has absorbed from that light those colors that the gas would radiate itself). Fraunhofer came closest to realizing these implications when he noticed that two dark lines in the Sun's spectrum coincided with two bright lines in the spectrum of his sodium lamp.

Fraunhofer's remarkable career was cut short when he died of tuberculosis at age thirty-nine, soon after being knighted and relieved of his taxes as a citizen of Munich.

In 1849, W. A. Miller at King's College, London, and Léon

Foucault in Paris independently recognized that the pair of lines Fraunhofer had found in the spectrum of the Sun—and that had matched the two lines in the spectrum of his sodium lamp—were present in the spectrum of sodium in the laboratory. Ten years later, Gustav Kirchhoff and Robert Bunsen were able to explain the full significance of that discovery: The Sun is an incandescent body surrounded by a gaseous atmosphere with a lower temperature. The lines mean that sodium is present in the atmosphere of the Sun. In the early 1860s, Miller and William Huggins, a British astronomer whose wealth allowed him to construct a private observatory, found that light coming from other stars had the same spectral lines as light from the Sun. A new era of astronomy had begun. From 1863 to 1868, Father Secchi at the Vatican provided the foundations of the classification of stars by the patterns of their spectra. Also in the 1860s Huggins in England and Henry Draper in the United States began to have success photographing spectra.

Though for a while it looked as though stars are all made up of more or less the same mixture of chemical elements, and expectations that they might be sorted into families by their spectra dwindled, further investigation showed that certain patterns of lines in the spectra are not alike for all stars after all. Hopes rising again, astronomers began to try to categorize stars. With luck, a few members of some family whose spectrum made the family members recognizable would be near enough to Earth so that their distances could be measured by parallax. This sample would reveal whether all members of that family had the same absolute magnitude and whether they could be used as distance calibrators.

In the 1840s, there was another development that would have dramatic long-range benefits for astronomy and astrophysics. Austrian physicist Christian Doppler discovered what is now called the Doppler effect. In everyday life we experience the effect with sound rather than light. We hear the pitch of a fire engine siren become lower as the vehicle passes and moves away. That's because as the fire engine approaches, sound waves coming to us from it are bunched up, shortened. As the engine moves away from us, sound waves from it are stretched out. In either case, the sound waves are "shifted" from the length they would be if the source were standing still. Our ears interpret the lengths of sound waves as different pitches; the longer the wave, the lower the pitch.

Doppler thought the same effect would occur with light, because light also can be thought of in terms of waves and wavelengths (review figure 4.6). Why shouldn't those be stretched or bunched up in the same way sound waves are? Just as our ears interpret sound waves of different lengths as different pitches, our eyes interpret light waves of different lengths as different colors—the longer the light waves, the nearer the "red" end of the spectrum. It would seem to follow that light from an object approaching us would be "blueshifted," because the waves would be shortened, and light from an object receding from us would be "redshifted," because the waves would be lengthened. In 1848, French physicist Armand Fizeau was the first to describe successfully redshift and blueshift in light. Today we use the term *Doppler shift* not only for sound but also when referring to all forms of light, that is, all radiation within the electromagnetic spectrum: gamma rays, X rays, ultraviolet rays, visible light, infrared, microwaves, and radio waves. (See figure 4.6.)

Doppler at first believed that red- and blueshifts were causing the differences of color observed in some binary stars, but other researchers, including Fizeau, soon pointed out that such a shift in starlight can't be seen at all as a visible difference in color. Instead, the shift shows up as a slight yet measurable shift of the spectral lines in light coming from a star or other source—lines such as those shown in figures 4.7 and 4.8. The shift is in the positions in the spectrum at which such lines appear—a shift either toward the "red" or the "blue" end of the spectrum. If the pattern has shifted toward the red end of the spectrum, the star from which that light is coming is moving away. If it has shifted toward the blue end, the star is approaching.

The light coming toward Earth from a star travels, of course, at the speed of light, regardless of what speed the star itself is traveling or whether it's coming closer or moving farther away. The blue- or redshift isn't a shift in the speed of light. It indicates instead that the light waves are being bunched up or stretched out (as the sound waves were with the fire engine). The amount of the bunching up or stretching out is directly related to the fraction of the speed of light at which the star (or other light source) is approaching or receding. In other words, the amount of red- or blueshift reveals how rapidly the star is approaching or receding. Huggins was the first to calculate the velocity of recession of a star from the shift of its lines.

Though the Doppler shift was obviously a remarkable help in plotting the movement of stars, astronomers also realized that a star's motion is likely to be more complicated than the simple approach or recession velocity the shift reveals. The Doppler shift by itself is only the measure of how rapidly a star is increasing or decreasing its distance from the Earth, not what direction it's taking or how fast it's actually traveling. Most stars don't move directly along our line of sight (directly away from or toward us), and so their recession/approach rate is not the same as their "total" velocity. For example, a star moving directly *across* (perpendicular to) our line of sight isn't increasing or decreasing its distance and shows no shift at all, yet that does not mean it is standing still. (See figure 4.9.)

Was there any way the Doppler shift could help in the calculation of a star's "total" motion? Perhaps even help reveal its distance? What about combining knowledge of the recession/approach rate with knowledge of the star's proper motion (which is movement across our line of sight)? Here was the problem: It was indeed possi-

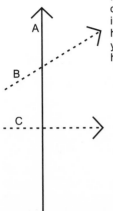

Imagine that you are standing on the Earth watching an object that you know is moving straight up. If you find it's increasing its distance from you at a rate of 200 miles per hour, moving directly away from you as shown in Line A, you know that it is traveling at a velocity of 200 miles per hour. 200 miles per hour is its "total" velocity.

But if its motion includes motion across your line of sight (Line B)—and it STILL is increasing its distance from you at a rate of 200 miles per hour, you know it has to be traveling faster than 200 miles per hour in order to achieve that recession rate. How much faster? That depends on the angle at which its path is tipped in relation to your line of sight. If it is tipped at precisely a 90 degree angle (Line C), there is NO speed at which it could possibly travel to achieve a recession rate of 200 miles per hour, indeed to achieve any recession rate at all. Its total velocity is across (perpendicular to) your line of sight.

Figure 4.9 The moving cluster method gives astronomers a way to calculate the angle at which the path of a cluster of stars is tipped in relation to our line of sight as viewers on Earth. The angle tells them what proportion of the stars' motion is represented by the recession/approach rate and what is perpendicular to our line of sight.

ble to determine a star's apparent motion across our line of sight— its proper motion—measured in arcseconds. However, that was not as helpful as it might seem, for an arcsecond, unlike a mile or a kilometer or a light-year, isn't an absolute distance and can't be transformed into one without knowledge of how far away the star is.

One ingenious method, known as the "moving cluster method," did use the understanding of Doppler shifts combined with other knowledge about the movement of stars to discover the total motion of *groups* of stars, and to measure their distances. It proved so effective that some of the measurements it provided weren't bettered until well after the advent of space-based astronomy. The technique is in fact still employed for more distant groups. The challenge was to calculate in what proportion two measurements (recession/approach rate and rate of travel across our line of sight) combine in the star's total motion. Here is where the advantage of having a group of stars, rather than a single star, came in. If a cluster of stars is moving through space, it's reasonable to assume that the stars are all traveling in nearly parallel lines and that the effect of perspective will cause these lines to seem to draw together toward a single point in the sky. You see a similar effect if you watch the two parallel rails of a railroad track from the last car in the train or from the locomotive. The rails seem to meet in the distance. For a group of stars moving away from or toward the Earth, the effect shows up if astronomers watch them over a period of years and plot their motion against the celestial sphere. The stars look as though they're getting closer together or moving farther apart, and the paths they are following seem to meet. In other words, the paths appear to be converging toward or diverging from a point.

By studying the pattern of this convergence or divergence—the way a star cluster seems to shrink or swell over time, and the location of the point where the stars' paths meet—researchers can determine whether a cluster is moving directly along our line of sight, or if not, at what angle to our line of sight it's traveling. (Again, see figure 4.9.) That angle tells them what proportion of the stars' total motion shows up in the recession/approach rate. Knowing that proportion, and what the recession/approach rate *is*, makes it possible to calculate a number, stated this time in miles or kilometers per second, for the stars' motion across (perpendicular to) our line of sight. It then becomes an answerable question of mathematics: How far away does a group of stars have to be for *this* velocity to produce *this* many arcseconds of shift across the sky in one year?

The moving cluster method provided the distance to the nearest star cluster, the Hyades, which forms the head of Taurus, the bull. The Hyades lie some forty parsecs away (approximately 140 light-years), about ten parsecs farther out than the parallax method alone could measure in the late nineteenth century. The Hyades include many kinds of stars. Knowledge of their distance allowed experts to calculate the absolute magnitude of stars in the cluster whose spectral lines identified them with particular families of stars. From there, the technique of comparing the apparent magnitude of stars belonging to the same family sufficed to reveal approximate distances to stars and clusters much farther away.

"Statistical parallax" is another technique that became possible with the discovery of how to measure the redshift of a star. It contributed directly to one of the first major advances in measurement that took place after the turn of the century. Astronomers choose a large group of stars, basing their choice on some common characteristic such as color or spectrum, and measure the stars' red- or blueshifts to learn their approach/recession velocities. Then they find the average of these velocities for all the stars in the group. The assumption, a reasonable one, is that stars are moving every which way in the sky at many speeds and in many directions, and that in a large enough group of stars all this motion averages out. So it also seems safe to assume that the average velocity along our line of sight is also the average velocity *across* our line of sight. Again, we are prepared to ask the question: How far away do stars have to be for *this* velocity to produce *this* many arcseconds of shift across the sky in one year? The result now is the *average* distance for the stars in the chosen group. That might seem to reveal very little. However, with some refinements, this method would later provide distances to some Cepheid variable stars, and these would be the bridge to objects many, many light-years farther away.

Along with increased ability to measure the distances to stars and groups of stars went a deepening curiosity about the larger picture—how everything fits together. What had all this burgeoning knowledge been leading men and women to believe about the overall shape, structure, and size of the universe? To answer that question, we must return to the eighteenth century.

CHAPTER

Upscale Architecture

{1750-1958}

&&The subject of the Construction of the Heavens, on which I have so lately ventured to deliver my thoughts to this Society, is of so extensive and important a nature, that we cannot exert too much attention in our endeavors to throw all possible light upon it. &&
William Herschel, 1785

In the decades following Galileo's observation that the streak of light called the Milky Way consists of myriad stars, few astronomers followed up on that discovery. There was plenty to keep them occupied much nearer home, within the solar system. Not until the middle of the eighteenth century is there a record of anyone suggesting any sort of "structure" in which all those stars—and perhaps the solar system as well—might be arranged. *Galaxias* is the Greek word for the Milky Way in the night sky, but the word "galaxy" didn't take on its modern meaning until the very late nineteenth and early twentieth centuries. Only then did clear evidence emerge that the

Milky Way star system—"our" star system—is a structure that is distinct and separate from many others like it in the universe. Put yourself in the place of those earlier astronomers. Imagine seeing only what they were able to see with the first telescopes and you will realize that such a concept, though not ridiculous, was simply unlikely to occur to them as a matter for scientific investigation or even meaningful speculation. There was no evidence arguing either for or against it. The subject fell more into the realm of fantasy—something the French court, in fact, found an interesting topic of conversation.

Thomas Wright, who called himself a philosopher, not an astronomer, was one of the first seriously to propose a larger arrangement. Wright was born in Durham, England, in 1711 and spent most of his teenage years, while apprenticed to a clock maker, immersed in astronomy. His father found that so distasteful that he burned young Thomas's astronomy books. Later, Wright became a sailor (giving that up after a storm on his first voyage), a mathematics tutor (until he became involved in a scandal with a clergyman's daughter), a navigation instructor for sailors, a surveyor, a successful teacher and consultant among the aristocracy on the subjects of philosophy and mathematics, and finally an author. All this he managed to do in spite of having a speech impediment.

In a book titled *An Original Theory or New Hypothesis of the Universe*, published in 1750, Wright suggested that the Milky Way, which he called the "universe" or the "creation," is a slab of stars whose center is a supernatural source of energy, goodness, morality, and wisdom. He thought that this slab might be one among many "creations" of its kind and that faint clouds of light, known as the nebulae, might be the other "creations."

The great east Prussian philosopher Immanuel Kant, who was a trained mathematician, read an account of Wright's ideas in a newspaper and was sufficiently taken with them to try to give them

William Herschel.
(Yerkes Observatory)

a more mathematical and scientific footing. Kant was not an experimental scientist, nor did he observe the heavens through telescopes. Instead he ruminated about the meaning of the observations and discoveries others were making. He agreed with Wright that if the Milky Way is a flat slab of stars, other fuzzy patches in the sky must be similar flat slabs. Our slab must be one of many in an enormous universe.

No End in Sight

One man who owned a copy of Thomas Wright's book was William Herschel, a prominent professional musician and composer who had become enamored of astronomy as a child. Born in Hannover, Germany, in 1738, Herschel first visited England when he was nineteen as a member of the band of the Hannoverian Guard.

Within the year he had resigned his commission and moved to England to pursue a musical career. He became director of music for the city of Bath and organist at the Octagon Chapel. Herschel spent most of his time teaching music students, engaging artists for concerts, and composing. His works include twenty-four symphonies, seven violin concertos, two organ concertos, and "glees and catches" and anthems for his choirs. Yet history remembers Herschel as an astronomer, for at age thirty-five, during hours when others were sleeping and whenever his pupils went away on holiday, he turned again seriously to his boyhood hobby.

In 1781, scanning the heavens with a telescope along with his younger sister, Caroline, looking for faint, distant objects, Herschel discovered the planet Uranus. As a result, King George III of England appointed him his personal astronomer, with few duties other than to keep the royal family advised on the subject of astronomy and any new wonders in the skies, and occasionally to impress visiting dignitaries. In 1787, the king also granted Caroline a salary.

The Herschels moved to Datchet, near Windsor, so that William could perform his new duties more conveniently. To supplement his modest stipend (less than he earned as organist), he began to make and sell telescopes, and he also set about building bigger and better instruments for his own use. His telescopes were reflecting telescopes (see figure 4.5), a type that had fallen somewhat out of favor. Herschel's success gave them a new popularity. His largest reflector was a forty-foot instrument with a forty-eight-inch mirror. Its tube was five feet in diameter. Herschel, still the musician, held a small concert in the tube to celebrate its dedication.

Although the size of the "forty-foot" made it a curiosity and a source of pride for Herschel and for the king, it really was something of a liability. So ungainly and unsafe was this behemoth that King George's workmen were reluctant to help maneuver it. What was worse, Herschel found himself wasting precious time and good

Herschel's forty-foot telescope. *(Yerkes Observatory)*

viewing nights (rare in England) explaining and demonstrating this wonder to sightseers, who included the king and his family and royalty and astronomers from all over the world. "Come, let me show you the way to heaven," the king was overhead murmuring to the archbishop of Canterbury, as he ushered him toward the telescope.

Herschel ended up preferring his more manageable twenty-foot reflector, though the use of that, too, involved physical risks. His sister, Caroline, wrote:

> My brother began his series of sweeps when the instrument was yet in a very unfinished state, and my feelings were not very comfortable when every moment I was alarmed by a crack or fall, knowing him to be elevated fifteen feet or more on a temporary cross-beam instead of a safe gallery. The ladders had not even their braces at the bottom; and one night, in a very high wind, he had hardly touched the ground before the whole apparatus came down.

Just before Herschel's forty-foot telescope was dismantled, seventeen years after his death, his son, Sir John Herschel, composed a ballad, and he and his family went into the tube on New Year's Eve to sing it. The great telescope's life ended as it had begun, with a concert rather than an astronomical observation.

For William Herschel, building bigger telescopes was not primarily a matter of satisfying his ego or impressing the king. He wanted to learn how the universe is constructed on the largest scale. His was the first project to map the universe in three dimensions, something astronomers are still trying to do two hundred years later. Rather than tackle the whole at once, Herschel chose seven hundred different regions of the sky on which to concentrate his attention. In each of these he meticulously counted the stars of different brightnesses and cataloged all the double stars he could find, for he hoped to measure their annual parallaxes.

Herschel's map turned out to resemble rather remarkably the late-twentieth-century image of the Galaxy, which is particularly surprising because he worked from unsound assumptions. He chose to assume that all stars are evenly distributed and that through his telescope he was seeing all the way to the outermost regions of the star-filled universe, so that his star counts really were accurate indications of the total number of stars. Also, like Newton, he decided to assume that all stars have essentially the same absolute magnitude. He picked the star Sirius, the brightest star in the night sky, as his standard. That is, he proceeded on the assumption that if all stars were the same distance as Sirius they would all look just as bright as Sirius, so the extent to which they appear to differ in brightness from Sirius can be used as a gauge of their distances. Herschel can't be so easily excused as Newton for this error, because he must have been aware of a strong argument coming from his friend John Michell. Michell, a natural philosopher best remembered as the first to suggest the existence of "dark stars"—what are now called black holes—pointed out that the stars of the Pleiades

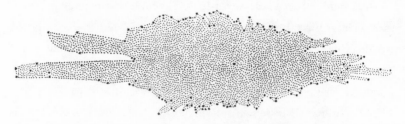

Figure 5.1 William Herschel's map—the "Grindstone"—was surprisingly like the modern picture of the Milky Way Galaxy.

definitely do not all appear equally bright. Yet, grouped as they are, they must surely be about equal in distance from Earth.

Herschel's map had the arrangement of stars as a spiky, flat, elongated blob, a model that got the nickname of the "grindstone," though it is difficult to imagine that any grindstone could be so ill-designed and jagged. (See figure 5.1.) The spiky edges were the result of dark rifts that Herschel observed in the Milky Way. He thought these were probably holes in space and that through them he was seeing emptiness beyond. To estimate the size of the grindstone, he decided to call the distance to Sirius, whatever it might turn out to be in miles or kilometers, one "siriometer." Herschel calculated that the grindstone measured 1,000 siriometers across and was 100 siriometers thick. When he made his estimate, the first detection of stellar parallax was still half a century in the future, but it's now possible to attach definite numbers to Herschel's scheme. Sirius turns out to be not quite 9 light-years away. That would make Herschel's grindstone about 9,000 light-years from end to end and 900 light-years thick. Modern estimates of the dimensions of the galaxy have it more than ten times as large.

William and Caroline Herschel also scrutinized the nebulae. The big question in the 1780s was: Are they clusters of many stars? The Herschels had the best equipment in the world for finding out.

William soon reported with delight that many nebulae *were* resolvable into stars. He even thought that he could *almost* make out individual stars in the Andromeda nebula, though astronomers now are sure that would have been impossible with his telescopes. In 1790, the Herschels confirmed the existence of another kind of nebula in which a cloud of luminous gas surrounded a single central star. Herschel thought this might be a planetary system in the making but not an independent cluster of stars similar to our own "grindstone." He also found that even with the best of his instruments he could not resolve all the nebulae into stars or find stars in them. Some must be clouds of gas.

Herschel gave up on his grindstone model later, when he discovered that many double stars are true binaries (two stars orbiting the same center of mass), with the pair of stars clearly the same distance from Earth yet differing in brightness. He was forced to conclude that stars do not all have the same absolute magnitude. At the same time, his larger telescope was revealing that beyond what he had previously thought were the farthest limits of the universe the stars go on and on. He could find no end to them. Such was Herschel's influence that other astronomers also abandoned his grindstone model. The question of structure on this scale didn't resurface in a significant way again until the middle of the nineteenth century, when another great amateur revived it.

William Parsons, the third earl of Rosse, who resided at Birr Castle in Ireland, was the feudal lord of the village of Parsonstown. He had been educated at Dublin and Oxford and served as a member of Parliament while still an undergraduate. In 1841, at age forty-one, he succeeded to the earldom, which gave him free time and an ample independent income to pursue his passion for astronomy. Lord Rosse also had a good knowledge of engineering and plenty of space to build a foundry and workshops. The lack of skilled labor in Parsonstown he soon remedied by training the laborers on his estate. He did not intend

The "Leviathan of Parsonstown." *(Royal Astronomical Society)*

to buy a telescope and install it at Birr Castle. He was going to design the thing, build it, and cast the mirror himself.

Though improvements in refracting telescopes had made them once again far more popular than reflectors both for observatories and private use, Lord Rosse's intention was to build a reflector larger than anything Herschel had used. Did he consider the weather in Ireland and wonder whether this was the most desirable location for the world's largest telescope? Later he wrote to his wife, "The weather here is still vexatious: but not absolutely repulsive."

Lord Rosse began experimenting with the construction of smaller telescopes and worked his way up. Eventually he achieved his goal: a tube fifty-six feet long and eight feet in diameter, with a six-foot mirror weighing four tons, set up to protrude like a giant cannon aimed at the sky from an amazing castlelike structure. This time there is no record of a concert. Professionals in the field of astronomy scoffed that Lord Rosse was more interested in designing and constructing telescopes than using them; however, he began observing with his "Leviathan of Parsonstown" in 1845, even before the supporting structure was completed. He aimed his celestial cannon at the nebulae.

Lord Rosse knew that these faint, fuzzy patches in the sky had both intrigued and frustrated William Herschel. Herschel, his sister, Caroline, and his son, John, a fine astronomer in his own right who spent some years at the Cape of Good Hope surveying far southern skies, had cataloged thousands of nebulae. Nevertheless, at the time Lord Rosse looked at them in the mid-nineteenth century, both with his smaller telescopes and with his "Leviathan," the nebulae were still one of the great enigmas of astronomy. In spite of the Herschels' observations, controversy continued over whether they were conglomerations of gas, perhaps not far away, which might be the birthplace of new stars and planets, or whether they were instead incredibly enormous clusters of stars, many too distant to be resolved by an earthly telescope.

Lord Rosse saw the nebulae as no one had before. By 1848, he had resolved fifty of them into stars. Some had complex structure, and, as Lord Rosse went on observing and drawing what he observed, more and more of them turned out to be spiral, lens-shaped formations. It became impossible not to suspect that the then out-of-date idea that these were star formations similar to our own, and extremely distant, might be correct after all. It also seemed likely that our own star system was, like them, spiral and lens-shaped, which was remarkably close to the way Herschel had pictured it in his "grindstone" model.

William Huggins, who like Lord Rosse had the wherewithal to build his own private observatory, and who had been one of the first to discover that light from the Sun and from other stars has similar spectral lines, also turned his attention to the nebulae. He analyzed the light coming from the Orion nebula, the Crab nebula, and oth-

The Orion nebula, studied by William Herschel, Lord Rosse, William Huggins, and many astronomers in the twentieth century, turned out not to be made up of stars as Andromeda is, but of gas lit by stars within it. It is in the Milky Way Galaxy, not a separate "island universe." *(National Optical Astronomy Observatories)*

ers, and found their spectra were like the spectra of hot, luminous gases, not the same sort of spectrum as light coming from the Sun and the stars. But he also discovered that light from other nebulae, the great Andromeda nebula for one, gave a continuous spectrum of the sort one would expect if it were made up of stars.

In 1885, the distant heavens gave earthly astronomers a spectacular opportunity, or at least many of them thought that was what it was. They had already judged that the Andromeda nebula, one of the largest of the spirals, was probably the closest. In this nebula a new star suddenly appeared and became bright enough to be just visible to the naked eye. Astronomers knew of only one kind of exploding star—a nova. Comparison of this star's brightness with the brightness of previous novae, and later with a nova in 1901, seemed to indicate that the Andromeda nova was relatively close to Earth. Of course if it was, that would mean that the whole Andromeda nebula must be close, by some estimates the nearest thing outside the solar system—certainly not a distant formation as large as the Milky Way.

The Capture of Light

While this study and speculation was going on, in the last quarter of the nineteenth century, astronomers were beginning to realize the potential value of a fabulous new tool: photography. Back in 1839, Louis Jacques Mandé Daguerre in France and Fox Talbot in England had almost simultaneously announced their discoveries of the photographic process. That same year, William Herschel's son, John Herschel, took one of the earliest photographs, a view of his father's forty-foot telescope through the window of his house at Slough.

Though there were some fine astronomical photographs in the mid-1800s, exposure times were not yet fast enough for photography to be of practical, routine use to astronomers. John Herschel's exposure time for the photo of the telescope was two hours. When Lord Rosse recorded his observations, he did so with drawings, not photographs. But when in the 1870s the use of dry gelatin plates reduced the exposure time required in terrestrial photography to about one-fifteenth of a second, a new epoch in astronomy began. It was no longer necessary to rely on words or drawings to share observations, or on memory to compare what a portion of sky had looked like on one night with is appearance on another. Photographs taken on successive nights or over a span of days, weeks, and years allowed astronomers to study how the sky changed. Photographic records took the place of such descriptions as Galileo's of Jupiter's moons, or

John Herschel's of the star Alpha Hydrae, in 1838, as he was sailing back to England from the Cape of Good Hope:

21 MARCH *Alpha Hydrae inferior to Delta Canis Majoris, brighter than Delta Argus and Gamma Leonis.*

7 MAY *Alpha Hydrae fainter than Beta Aurigae, very obviously fainter than Gamma Leonis, Polaris or Beta Ursae Minoris.*

10 MAY *Alpha Hydrae much inferior to Gamma Leonis, rather inferior to Beta Aurigae. It is still about its minimum.*

11 MAY *Alpha Hydrae brighter than Beta Aurigae no doubt.*

12 MAY *Castor and Alpha Hydrae nearly equal.*

"Very obviously fainter" . . . "rather inferior" . . . "much inferior" . . . "brighter, no doubt" . . . "nearly equal." What a difference photography was to make in the precision with which observations could be reported and compared!

The English astronomer Isaac Roberts pioneered long-exposure photographs, allowing more light to enter the camera through the telescope than would happen in a quicker exposure. In 1888, he was the first to take photographs of Andromeda. Roberts's photographs confirmed that Andromeda is spiral in shape, and clearly revealed the spiral arms in the galaxy's outer regions. However, even those photographs couldn't settle the question of what Andromeda actually *is.* Eleven years later, photography was first put to the task of recording a spectrogram of Andromeda, which indicated that it was a "cluster of sun-like stars." Yet Huggins had just previously seen mixed dark and light bands from Andromeda!

The confusion about the nebulae was part of a continuing debate about the larger picture. Friedrich von Struve, who had first measured the parallax of Vega, believed that the Milky Way disk's edges extend to infinity, with interstellar matter absorbing the light from remote regions so that they remain eternally hidden from us. Others argued the pros and cons of a proposal that the Milky Way consists of concentric rings of stars.

Partly because of the advent of photography, increasing attention was given not only to the position of stars but to their motions. It was soon confirmed that stars are more likely to move in the plane of the Milky Way. In 1904, Dutch astronomer Jacobus Cornelius Kapteyn discovered that the majority of those stars that are easiest to observe move in two streams toward different parts of the sky. Like William Herschel, he counted stars and found that Herschel had not been far off in his conclusions about their distribution.

Kapteyn thought the Sun was near the center of the Milky Way system; American astronomer Harlow Shapley would soon argue, based on his study of globular clusters, that it was not. In 1913, the Dutch astronomer C. Easton concluded that the whole universe was one large spiral, shaped like the spiral nebulae. He thought that these nebulae were only miniatures of the greater spiral, within its boundaries.

The definitive answer was still almost a quarter century away. However, at the turn of the twentieth century, astronomy was not far from a tremendous breakthrough when it came to establishing a foothold beyond the range of parallax measurement. Astronomers had been having some success putting together the ingredients necessary for such an advance: the ability to identify classes or families of stars by some characteristic that would not change with distance and the ability to measure the distance to at least a few of the stars in a family—or, failing that, at least to decide that a number of stars in the family were all approximately the same distance from Earth.

Astronomers were using the parallax method to measure the distances to as many stars as possible within parallax range and making catalogs of these stars and distances. They were continuing the search for distinguishing characteristics—color perhaps, or some pattern of variation of color or brightness, or patterns of spectral lines. And they were attempting to identify groupings of stars that are all approximately the same distance from Earth. With this combination of efforts, researchers were managing to edge themselves farther and farther into the cosmos.

There was a risk, just as there was in the analogy with the elephants and the giraffes. Discovering a weakness in one rung of the cosmic distance ladder could, and would several times, necessitate recalibrating the entire structure. But this was a problem astronomers had learned to live with, making repeated adjustments, hoping their margins of error were not greater than they estimated and that somewhere along the line there would be independent evidence to show that their measurements hadn't been far from the mark. Things were going fairly well. They were about to get much better.

The Cepheid Yardstick

In the early to mid-nineteenth century, the fashion for observatory building had spread to the United States. At first most of the telescopes there were imported from Europe. In the 1830s, there were good refracting telescopes at Yale University and at Wesleyan University in Middletown, Connecticut, but no actual "observatories" at either place. Astronomers at Yale just stuck their telescope out a window. In 1838, Williams College, in Williamstown, Massachusetts, opened its Hopkins Observatory, housing a ten-foot Herschel reflector bought by Professor Albert Hopkins in England. The Harvard College Observatory was founded in 1839 but didn't have a building or very much in the way of equipment. Earlier, in 1815, a delegation had gone to England to purchase a telescope for Harvard, found the desirable instruments too expensive, and come back

Henrietta Swan Leavitt.
(Harvard Observatory)

empty-handed. In 1843, the local citizens of Cambridge, Massachu-setts, disgruntled that there was no telescope available through which they could view the great comet of that year, offered to share with the university the cost of purchasing one. Harvard ac-cepted, and the telescope was acquired from a distinguished firm in Germany.

By the 1890s, the Harvard College Observatory had become a world-class institution. It was there that new rungs on the cosmic distance ladder were about to be nailed into place.

Henrietta Swan Leavitt was born in 1868 and studied at what would later become Radcliffe College in Boston, then known as the Society for the Collegiate Instruction of Women. At the nearby Harvard College Observatory, the eminent astronomer Edward Pickering was cataloging and analyzing stars and mentoring younger scholars. There were few if any women among these budding as-tronomers, though women were hired to do the painstaking drudg-ery of writing down in endless rows of figures the positions and brightnesses of stars. However, Pickering occasionally encouraged someone from among his volunteer or underpaid female clerical staff to take on a more creative and challenging assignment. In 1895, Henrietta Leavitt became a member of Pickering's staff. She started as a volunteer, received a permanent paid position in 1902, and soon became head of a department.

In 1908, Leavitt was looking for stars that vary in brightness, hoping to find a group of them that were all approximately the same distance from Earth. It was logical to assume that all the stars in one of the Magellanic Clouds are, by cosmic standards, approximately the same distance away.

The Magellanic Clouds are two star formations that are not visible at any time of year in most of the Northern Hemisphere, where they never rise above the horizon. Seen from the Southern Hemisphere, they are large misty smudges of light that could be mistaken for thin, veil-like clouds faintly lit by the Moon on a fair night. The Australian aborigines believed that the Large Cloud was a part of the Milky Way that had been torn away. Europeans knew of the existence of these clouds before Ferdinand Magellan's voyage around South America in 1521 and called them the Cape Clouds. But Magellan's official recorder, Antonio Pigafetta, suggested that they be renamed the Clouds of Magellan in honor of that great explorer, who died just short of completing his circumnavigation of the globe.

Astronomers didn't pay the clouds much attention until John Herschel studied the Southern Hemisphere skies in the 1830s and hypothesized that these clouds were fragments detached from the Milky Way. This was not plagiarism from the Australian aborigines. Herschel was in South Africa, not Australia. The younger Herschel thought this might mean the Milky Way was breaking up and that his father had been right to speculate that it couldn't last forever . . . indeed, that the past might not be infinite either.

By the end of the century, there was general agreement that the Magellanic Clouds were made up of stars. There was less consensus about whether the clouds are part of the Milky Way system or distinct from it but closely related. It was Pickering's cataloging, plus the techniques for measuring distances to groups of stars, that began to provide a clearer understanding of the Magellanic Clouds as the new century began. Today, astronomers measure their distance from Earth at about 169,000 light-years and consider them satellite galaxies of the Milky Way, but the earlier name has stuck— the Large and Small Magellanic *Clouds*. When Leavitt examined them, their distance was still unknown.

The reasoning behind Leavitt's study was that if stars in the Magellanic Clouds are all approximately the same distance from Earth, it is not differences in distance that cause some of them to look brighter than others. It seemed safe to conclude that stars that look bright there really do have greater absolute magnitude than stars that look dim there, and meaningful comparisons could be made. The Magellanic Clouds were close enough for individual stars to be identified and studied, though not close enough for their distances to be measured by direct parallax.

In the closing years of the nineteenth century and the early years of the twentieth, the Harvard College Observatory had an outpost known as it Southern Station in Arequipa, Peru. By then, photography had come into wide use in astronomy, not only allowing researchers to compare observations much more systematically but also making it possible for important discoveries to be made away from the telescope itself. From plates taken in Arequipa, Leavitt, in Boston, was able to identify 2,400 variable stars in the Small Magellanic Cloud.

Leavitt found that some of these variables had a remarkably dependable pattern to their variation: a steep increase to maximum brightness and then a more gradual falloff in brightness. Some took much longer than others to complete the pattern, and their range of brightness was also different. Nevertheless, the overall pattern was recognizable enough to set them apart. Leavitt noticed that among her sample in the Small Magellanic Cloud, the brighter a star of this type was, the longer it took to complete the pattern. The brightest ones took almost a month (some are now known to take over three months), the faintest only a day or so.

This relationship between a star's period (the length of time it takes a star to complete the cycle) and its brightness was precisely the sort of clue for which Leavitt and others had been searching. The period of pulsation was a characteristic that wouldn't change with distance. Leavitt found twenty-five stars of this distinctive family in the Small Magellanic Cloud, all of which showed a clear relationship between brightness and period. She compared these "light curves" with those of previously discovered variable stars nearer to Earth, in the Milky Way Galaxy, and found a match in the star Delta Cephei. Cepheids is hence the name given these variable stars.

Leavitt published her initial findings in 1908. Four years later, in 1912, she had compiled enough evidence to show that Cepheids could potentially provide a much more reliable way than any known before to pin down distances both in the Galaxy and far beyond it.

Wasn't it working backward to have a discovery about stars so far away lead to a method for measuring distances closer to home? Not really. Studying the stars in the Galaxy and comparing their distances can be an extremely confusing undertaking. They are certainly not all the same distance from Earth, and brightness is no gauge of their distance. Take two hypothetical stars, Star A and Star B. Say that Star A's absolute magnitude is twice as great as Star B's—Star A is a *much* brighter star. Nevertheless, if Star A is twice as far away from Earth as Star B, Star A will actually look the fainter of the two to observers on Earth. (Recall the inverse square law and the discussion of the two lightbulbs.) The key to Leavitt's discovery was finding a family of stars with a good sample of its members in an area where she knew all the stars were approximately the same distance from Earth. The Magellanic Clouds were such an area, and there was no area like that within the Milky Way.

It might seem, however, that the Magellanic Clouds are surely large enough so that differences in stars' apparent magnitudes might be deceptive, just as they are in the Milky Way Galaxy. However, referring to the hypothetical situation above, *no* star in a Magellanic Cloud is twice as far away from Earth as another in the same Cloud. Think of it this way: From my house here in the eastern United States, I can say that my fence is twice as far away from me as my garage. But if I were in Tokyo, I could not say that. For all intents and purposes, from Tokyo this fence and garage are the same distance away. So it is with the Magellanic Clouds. The stars there are so far away that astronomers can treat them all as being the same distance from Earth.

Leavitt had found in the Small Magellanic Cloud that Cepheids with the same range of absolute magnitude had the same period of variation, which meant that if she knew how one Cepheid's period related to another Cepheid's period, she also knew the relationship between their absolute magnitudes. For example, Leavitt found in her sample that if one Cepheid had a period of three days, and another had a period of thirty days, the second was six times brighter than the first. This meant that anywhere else in the sky she discovered a Cepheid variable star, she could measure its period of variation and be fairly sure that would tell her how bright that star would appear *if* it were sitting among the stars in the Small Magellanic Cloud, and how its actual distance compared with stars there. This was definitely an advance in measuring stellar distances, but, again, as with Kepler's third law and Herschel's siriometers, what Leavitt had was a system of *relationships*, not absolute distance measurements. No one then knew the exact distance to the Magellanic

Harlow Shapley.
(Harvard Observatory)

Clouds, or the distance to or absolute magnitude of any Cepheid in the Milky Way Galaxy. No Cepheid—not even Polaris, the North Star, the nearest Cepheid—was close enough to be measured by the parallax method. The cosmic distance ladder had been extended by many rungs, but it didn't reach the ground!

Not long after Leavitt's discovery, the Danish astronomer Ejnar Hertzsprung sought to remedy this situation. He applied a variation of the statistical parallax method and estimated the distance to two Cepheids. Using the relationships between period and absolute magnitude that Leavitt had established, he went on to calculate that the Small Magellanic Cloud is 30,000 light-years away. That distance, though far short of the nearly 169,000 light-years astronomers currently measure it to be, was much greater than anyone had been expecting.

Leavitt's "standard candles," as she dubbed the Cepheid variables, almost immediately drew the attention of other astronomers, and one of them was Harlow Shapley. Shapley, born in Missouri in

1885, had come to astronomy by a serendipitous path. As a teenager he'd had a job as a crime reporter for a newspaper in Kansas, and he'd thought his formal education would be in journalism. Shapley arrived at the University of Missouri in 1907, only to find that the school of journalism hadn't been built yet. When he returned a year later, nothing had changed, and Shapley was out of money and patience. As he later described it, he was "all dressed up for a university education and nowhere to go." He opened the university catalog and began reading about the courses, starting at the front with the letter A. Archaeology didn't appeal to him. On the next page was astronomy.

Four years later, with both a B.A. and an M.A. from Missouri, Shapley moved on to Princeton, where he began to investigate a different kind of variable from the Cepheids Leavitt had been studying. He concentrated on systems called "eclipsing binaries" in which a pair of stars orbit one another (or, to be more exact, orbit their common center of mass) with one star periodically eclipsing the other. The effect is that of a variable, for the light output changes as the stars eclipse one another. Shapley wanted to find out whether Cepheid variables might actually be binaries.

In 1914, Shapley's work came to the attention of George Ellery Hale, at the Mount Wilson Observatory in California. Hale had been the force behind the construction of the observatory, and at his invitation Shapley came to Mount Wilson to work with the sixty-inch reflecting telescope there ("sixty-inch" refers to the size of the mirror), the largest telescope in the world at the time, for Lord Rosse's had fallen into disrepair. Soon Shapley was able to satisfy himself that Cepheid variables are not binaries. He discovered that they actually swell and shrink, the result of instability below their surfaces.

Astronomers now know that Cepheids are elderly stars that have passed through their "main sequence" phase (the lengthy phase during which a star steadily converts its hydrogen to helium) and become "red giants." When the main sequence phase ends, the star starts to shrink. As it does, it heats up, and the heat flows into the outer layers of the star, energizing atoms of singly ionized helium. *Singly ionized* means there is an electron missing from these atoms. The increase in energy knocks off yet another electron, leaving the atoms "doubly ionized." Doubly ionized atoms readily absorb light. As a result of that absorption, the atmosphere of the star becomes opaque, holds in heat like a thermal blanket, gets hotter still, and expands. The star's outer layers swell, billowing out to about a hundred times the star's earlier size. As the star expands, it cools, for the energy has more room to spread out, and as the helium atoms cool they once again change from a doubly ionized state to a singly ionized state. The atmosphere becomes transparent again and begins to shrink. The cycle starts over. That is the mechanism behind the pulsation of a Cepheid. It swells and shrinks and changes brightness repeatedly at a steady rate.

Shapley, like Hertzsprung, decided to try to measure actual distances to Cepheids, using a method of his own. The Sun is not a Cepheid variable, but knowing something about the reasons for their pulsation (though not everything described in the previous paragraph) gave Shapley a new way to compare a Cepheid with the Sun. Having the distance and size of the Sun, he could then calculate a Cepheid's absolute magnitude. Cepheids turned out to be some of the brightest of all stars.

With his new understanding and the finest telescope in the world, Shapley began looking for a rich source of Cepheids and found it—the unattractively named globular clusters. These are dense, jewel-like, spherical clusters of stars. Using the Cepheids in them as standard candles, Shapley began calculating their distances.

He made three discoveries about the globular clusters that he thought significant. First, in all the globular clusters to which he

A globular cluster—M92 NGC6341, in Hercules.
(National Optical Astronomy Observatories)

could calculate the distances, the brightest stars were always approximately the same absolute magnitude. Shapley had a new yardstick.
Even if he didn't see a Cepheid in a cluster, he could assume that
the absolute magnitude of the brightest stars there was the same as
that for the brightest stars in clusters he had already been able to
measure. Shapley's second discovery about the globular clusters was
that they appeared to be spread equally above and below the plane
of the Milky Way, and this seemed to indicate that they are part of
the same system. Third, some of them were distributed much farther out than the stars of the Milky Way—so far out, in fact, that
later in the century there would be confusion as to whether some
groupings of stars were globular clusters in the Galaxy's halo or
separate dwarf galaxies.

It occurred to Shapley that globular clusters might form a skeletal outline to the Milky Way. Relative to the solar system, the distribution of the clusters appeared to be lopsided. A great many of them were gathered in an enormous sphere, a cluster of clusters far from the solar system, and centered on a point in the direction of the constellation Sagittarius. Shapley published his results in 1918 and 1919, interpreting them to mean that the center of this spherical group of globular clusters was the center of the Milky Way system. The Sun was nowhere near that center.

You would expect a clamor of outrage and argument to follow such an announcement. Humanity evicted once again—forced to move down yet another notch from that original exalted position in the center of the universe! It's surprising to learn that there was little public or scholarly outcry about Shapley's discovery, though there were conflicting theories at the time. What did shock astronomers and provoke a great deal of controversy was Shapley's measurement of the size of the Milky Way system. It was so huge that Shapley surmised that the system and its outline of globular clusters—in all, by his calculation, 300,000 light-years across—must indeed be all there was to the whole universe. Nebulae such as Andromeda were not independent systems. At most they were minor satellites of the Milky Way. Experts now know that Shapley overestimated the size of the Galaxy because he failed to take into account the way intervening dust affects light coming from the globular clusters. The dust dims the light, and the sources look fainter and farther away than they really are. Astronomers now measure the Galaxy at approximately 100,000 light-years, only a third as large as Shapley thought it was.

There was opposition to Shapley's estimate from Heber Curtis of the Lick Observatory in Santa Cruz, California. His road to astronomy had been almost as odd as Shapley's. Curtis had been a professor of classics and Latin, but when the college at which he taught merged with the University of the Pacific, he changed hats and became a professor of astronomy and mathematics. Curtis wouldn't accept that Cepheids could be used as reliable standard candles. He insisted that the spiral nebulae were other systems far outside the Milky Way, that the Milky Way was much smaller than Shapley's estimate, and that the solar system was the center of the Milky Way. Curtis suggested that the 1885 nova in Andromeda, whose apparent magnitude had been taken as indication that the Andromeda nebula was well within the Milky Way system, might actually have been of much greater absolute magnitude than the nova in 1901 to which it was being compared—and hence much farther away than most astronomers thought.

Shapley scoffed at this suggestion. If it was correct, and the Andromeda nebula was another star system like the Milky Way, not nearby at all, the 1885 nova in Andromeda would have had to have an absolute magnitude equal to a billion ordinary stars, in order to look as bright as it did. Shapley thought it was absurd to imagine it could have been that bright, or that the universe could be large enough to contain a great many independent systems as big as the Milky Way. Furthermore, if nebulae like Andromeda were rotating at the rate some observations indicated, and were as far away as Curtis was claiming, they had to be rotating at a speed faster than the speed of light. That was impossible.

Curtis, undiscouraged, proceeded to search for other novae in Andromeda, so as to have more to compare with the 1885 nova. He discovered several, all much dimmer than the nova of 1885, and that added weight to his argument that the 1885 nova was indeed exceptionally bright. He insisted that it was the dimmer and much more common novae, not the uncharacteristically bright one, that should be compared with novae elsewhere in order to calculate the distance to Andromeda. Curtis decided that Andromeda was hundreds of thousands of light-years away, far outside the Milky Way. Shapley still disagreed.

In 1920, the arguments between Shapley and Curtis culminated in a debate arranged by the National Academy of Sciences in Washington, D.C.—attended by Albert Einstein, for one. The debate settled nothing. Hindsight shows that each man was right about some things and wrong about others. Together they could have put together a good picture of the universe. Most thought that Shapley came off the worse in the debate, though astronomers were soon agreeing that his location of the center of the Milky Way Galaxy was correct. Shapley left Mount Wilson not long after the debate to become the director of the Harvard College Observatory. Pickering, Henrietta Leavitt's mentor there, had died in 1919.

From Nebulae to Galaxies

The first two decades of the twentieth century ended without a resolution to the question of how large the Milky Way star system is and whether there is anything else besides it in the universe. The nebulae were still a puzzle. The "island universe" theory that some of the nebulae were other systems on the scale of the Milky Way was still in dispute even as late as the 1920s. The opposition had some observational data weighing in on its side. There was evidence that the nebulae were insignificant in size compared to the Milky Way—not its equals at all. Earlier indication that some nebulae had the spectra of stars (and must not therefore be made up only of gas) was called into question by the discovery that nebulae sometimes reflect light from elsewhere, so that not all the light seen from them is their own. There were also Shapley's arguments: The enormous size of the Galaxy made it highly improbable there should be others like it, and nothing could rotate faster than the speed of light. Waiting in part on the unraveling of these mysteries, an upheaval was in the making that would rival Copernicus's De Revolutionibus as a watershed in the history of astronomy.

Back near the beginning of the century, Vesto Melvin Slipher, a young midwesterner with only an undergraduate degree from Indiana University had come to work at the Lowell Observatory in Flagstaff, Arizona. Slipher was a quiet man, methodical and meticulous. When it came to insisting on tying up loose ends before announcing a discovery, he rivaled Copernicus. Slipher spent his entire career at Lowell. While working there, he earned his M.A. and Ph.D. from Indiana. In 1916, he became the observatory's acting director; in 1926, its director.

The wealthy astronomer Percival Lowell, who had built the observatory and was its director at the time he hired Slipher, subscribed to the belief that the nebulae could be other solar systems in an earlier stage of formation. He set Slipher to work measuring spectra of spiral nebulae, looking for Doppler shifts in their light.

This was no easy assignment. Slipher couldn't merely point the telescope at a nebula and snap the camera. Exposures lasted for twenty to forty hours, over several nights. A researcher couldn't stray far from the unheated telescope dome—purposely left cold because heat could mar the image—for he had to be certain the nebula remained at the center of the field of vision. The result at best was a faint, diffuse image of the nebula, not a concentrated point of light like a star. Using the spectroscope to spread it out farther made it even fainter. The spectral lines were hard to identify, sometimes too faint if the spread-out image was too large. On the other hand, an image sufficiently bright for the lines to stand out clearly was apt to be too small for the shift in them to be measured.

In spite of these obstacles, in 1912 Slipher succeeded in obtaining four spectrograms showing the Doppler shift in the light from the Andromeda nebula. All four showed the light to be blueshifted. In other words, the familiar patterns of absorption lines produced when light passes through a particular gas were shifted toward the blue end of the spectrum. (Refer to figures 4.6 and 4.7.) Doppler and Fizeau had shown that such a shift signifies that the light source is approaching, and Slipher accordingly interpreted the shift he observed to mean that the distance between Earth and the Andromeda nebula was growing smaller.

Between 1912 and 1914, Slipher painstakingly pushed his equipment to its limits and measured Doppler shifts of twelve more nebulae. Andromeda proved to be the exception. Only one other had a blueshift. The others were redshifted. Slipher calculated they were rushing away at speeds of hundreds of kilometers a second.

Slipher, characteristically, didn't jump to conclusions. Thirteen nebulae among all those known in the skies was too small a sample to claim that it indicated all or most nebulae are receding from Earth. However, in 1914, with great modesty, he reported his findings to the American Astronomical Society. John Miller, who had been one of Slipher's professors, described the event: "Something happened which I have never seen before or since at a scientific meeting. Everyone stood up and cheered."

Slipher went on designing and improving his own instruments and continued to find that most of the nebulae he was able to study did indeed show redshifts. In early 1921, he reported a nebula that according to his calculations was increasing its distance at a speed of approximately 2,000 kilometers per second. In 1922, he sent Arthur Eddington, an eminent physicist at Cambridge University in England, measurements for forty spiral nebulae, thirty-six of which were receding. Eddington was intrigued. He also was a cautious man, but he went so far as to suggest that this discovery about the nebulae, which were widely thought to be the most remote objects yet known, might be a hint about "general properties of the world," by which he meant "the universe." By 1925, astronomers had measured forty-five nebular Doppler shifts—Slipher forty-one of them, other astronomers the remaining four. The score was now forty-three redshifts to two blueshifts. What might have been a coincidence was definitely beginning to look like a trend.

Clearly, Slipher had made a discovery of enormous importance, but it wasn't obvious at first what it signified. Slipher's own initial interpretation was that the drift of the solar system through space was increasing the distance between it and the nebulae. One problem with interpreting the significance of the redshift was that knowing the nebulae were moving away from Earth, or Earth from them, still didn't answer how far away they were or what they were.

Now it so happened that when Slipher first announced his findings about redshifts to the American Astronomical Society in 1914, a young man named Edwin Hubble was in the audience. Hubble's background was unusual. His earlier university training hadn't been in astronomy at all. By the time he came to astronomy full-time, he was already a successful lawyer.

Born in Missouri in 1889, Hubble attended both high school and university in Chicago and then went to Oxford on a Rhodes scholarship. He was a polymath who was ready to tackle just about anything, including tank diving and high-level amateur boxing. But it was astronomy that won him in the end. He practiced law for only a few months before going back to the University of Chicago to study astronomy and work as a research assistant at the Yerkes Observatory. "All I want is astronomy," he said. "I would much rather be a second-rate astronomer than a first-rate lawyer." In 1917, with Ph.D. in hand and an offer of a job at Mount Wilson, Hubble first went off to France to fight in World War I and then returned in 1919, arriving at Mount Wilson just before Shapley left there to take up the post at Harvard.

With the 60-inch reflecting telescope at Mount Wilson, and sometimes the more powerful 100-inch reflector that had just come into service in 1918, Hubble began to investigate the nebulae. By 1922, he had confirmed that nebulae that do not have a spiral structure don't shine with their own light. Either they shine by light reflected from stars within or near them, or they absorb enough energy from nearby stars to cause the hot gas of which they are composed to glow. Hubble's study confirmed earlier strong suspicions that these nebulae are part of the Milky Way system, but that didn't settle the question of the spiral nebulae. Hubble turned his attention to those next.

He was sure that some of the spiral nebulae are made up of stars, but for many of them, even by using a magnifying glass to scrutinize the best photographs he was getting with the 100-inch telescope, he couldn't produce the sort of evidence that would make him, and others, certain this was so. Hubble decided to investigate a faint patch of light called NGC6822, which *could* be resolved into stars. And there, in 1923, he found some of those stars varying in brightness. At first Hubble failed to recognize their significance. He turned instead to the Andromeda nebula, where other astronomers had been discovering faint novae.

In the autumn of 1923, Hubble was using the 100-inch to photograph Andromeda night after night. His observations were part of a survey to search for novae there that might be used to test Curtis's ideas about nebulae. Hubble almost immediately found a couple of novae and another faint object that he at first thought was a third. At this point he decided to delve into the Mount Wilson archives to look for this star on older photographic plates. The comparison showed that it was actually a variable star, a Cepheid with a period of approximately a month, which meant that its absolute magnitude at its brightest was about 7,000 times as bright as the Sun. For it to appear as faint as it did, it had to be about 900,000 light-years away. Hubble looked again at the photographs he had recently taken of the nebula NGC6822, and this time he recognized the varying stars as Cepheids, allowing him to calculate that NGC6822 was about 700,000 light-years away.

These measurements to Andromeda and NGC6822 would later prove to be underestimates. Nevertheless, they settled the question whether the spiral nebulae are nearby or are remote, independent "island universes." By Hubble's measurement, the Andromeda nebula was much farther away than any star in the Milky Way system and definitely outside it. That indistinct, oval blur that we see in the night sky in the Northern Hemisphere was another system—a collection of millions, perhaps billions, of stars. Astronomers now measure its distance as about 2¹/₄ million light-years. The Andromeda galaxy, as we now call it, is the farthest object visible from Earth with the naked eye. At that distance, the nova of 1885 in Andromeda had to have been much brighter than any ordinary nova. It was in fact something rarer, a supernova, the explosion of a star at the end of its life, as bright as nearly a billion Suns.

Hubble rushed news of his discovery about Andromeda and NGC6822 to Harlow Shapley. On first reading Hubble's message, Shapley turned to his colleague Cecilia Payne-Gaposhkin and commented, "Here is the letter that has destroyed my Universe."

Hubble christened Andromeda and other similar independent systems "extragalactic nebulae." (He never called them galaxies.) By the end of the year, he had resolved the outer part of the Andromeda nebula into stars, and a year later, he was confident enough about the nature of the spiral nebulae in general to present his findings to the American Astronomical Society. His paper won an award donated by the American Association for the Advancement of Science to the two most outstanding and important papers presented at the meeting. The other winner was a paper on the digestive tracts of termites.

Over the next five years, Hubble went on accumulating evidence and began developing techniques for estimating the distances to galaxies (as we shall call them henceforth in this book in spite of the fact that Hubble did not) out beyond the range in which individual stars could be seen and identified as Cepheids. One technique resembled the one Shapley had employed when estimating the distances to the globular clusters: assuming that the brightest stars in all galaxies were approximately the same absolute magnitude and using these stars as distance indicators. That allowed Hubble to measure galaxies four times as remote as the farthest galaxy in which he could find a Cepheid. He estimated this range as about 10 million light-years.

Hubble didn't stop there. He began hammering new rungs into the cosmic distance ladder at a furious pace. He decided that globular clusters could be used as a standard, assuming that the brightest globular clusters in all galaxies are approximately the same absolute magnitude. To go farther yet, Hubble decided to assume that all galaxies have approximately the same absolute magnitude, or at least all fall within a narrow range. Recognizing that there would be some amount of error in this method, he nevertheless calculated that his distances would be no more than three times too large or three times too small. Hubble estimated that his techniques took him out to about 500 million light-years—a volume of space containing about 100 million galaxies. Others would later refine his method and measure even farther by assuming that the brightest galaxy in a cluster of galaxies has approximately the same absolute magnitude as the brightest galaxy in every other cluster of galaxies.

The accuracy of all this measurement stood or fell in great part on the reliability of the measurement of the distance to Cepheid variables. The ladder stood on that footing. The Cepheid yardstick has been revised several times over the years. Nevertheless, however rough the initial measurements were, Hubble did establish once and for all that the universe extends for billions of light-years. His results seemed to indicate that the distribution of galaxies and galaxy clusters is fairly uniform throughout space. Hubble's investigations and those of his successors into the nature of Andromeda and other galaxies also turned a mirror on our own Galaxy, helping astronomers get a better idea of what it must be like as a whole, for of course they couldn't see it from a distance in space as they could Andromeda.

In the years following Hubble's discovery that at least some of the nebulae are well outside the Milky Way, and the coinciding realization that our own system is one galaxy among many others, some astronomers still were reluctant to trust Cepheids as reliable distance calibrators. There was one particularly nagging problem. As more and more galaxies were measured, most of them turned out to be quite a bit smaller than the Milky Way. Andromeda was only one-sixth as large. Others were smaller still. Was the Milky Way Galaxy really exceptionally large? Perhaps by far the largest? That seemed suspicious enough for astronomers to have some doubts about the measurements.

By the 1940s, the bright lights of the rapidly growing city of Los Angeles were making Mount Wilson a less-than-ideal location for a telescope. However, in the middle of the decade, at the height of World War II, these lights were often blacked out because of the threat of bombing raids. Though the citizens of Los Angeles undoubtedly found this unpleasant, it was a stroke of luck for astronomer Walter Baade, who because he was a German national had not been allowed to join the war effort in military research, but instead was left in almost solitary splendor at Mount Wilson, with the 100-inch telescope virtually all to himself.

Born in Schröttinghausen, Germany, Baade had received his Ph.D. from the University of Göttingen in 1919. In 1931, with the political climate in Germany changing, he had come to the United States, where he spent twenty-seven years working at the Mount Wilson and Mount Palomar Observatories before returning eventually to his native country.

Under ideal viewing conditions during the blackouts, Baade studied the stars in the Andromeda galaxy through the 100-inch telescope. He found that those in the center of that galaxy and in what he called its "outer skeleton" or halo tended to be red and yellow, whereas those in the spiral arms were white and intense blue. Baade concluded that the stars must belong to two different "populations." He christened the white and blue stars Population I stars. These turn out to be hot young and middle-aged stars. The red and yellow ones he called Population II stars. These are much older stars.

Beginning in 1948, Baade was one of the first to use the 200-inch reflecting telescope (known as the Hale Telescope) on Mount Palomar, northeast of San Diego. He was puzzled to find Cepheid variables, with a difference, in the Andromeda galaxy halo. They were four times fainter than the Cepheids in the rest of the galaxy, the Cepheids Hubble had used to make his measurements. Baade, putting this finding together with what he had earlier discovered about the two different star populations, concluded that each population must have its own kind of Cepheid variable.

All Hubble and his associates had known were Population I Cepheids, the ones in the spiral arms of the Andromeda galaxy. The result was that they had been comparing apples and oranges, for the Cepheids they'd used for comparison were actually fainter Population II Cepheids. For this reason, Hubble's measurements of the distance and size of the Andromeda galaxy had been too small. Baade's discovery doubled the size and age of the universe. This increase helped clear up an embarrassing conundrum, for Hubble's measurements had made the universe younger than the Earth, whose age geologists had already calculated.

Baade's discovery also led to a far better understanding of spiral galaxies, including the Milky Way. As astronomers recalculated their measurements, they found that spirals are all much larger than Hubble had estimated, and far more remote. While the mental picture of the universe swelled in size and distance, the picture of the Galaxy shrank accordingly. Clearly, Shapley's 300,000 light-year measurement for the diameter of the Galaxy was an overestimate. The diameter was closer to 100,000 light-years. The Milky Way was not ten times larger than most spirals and six times larger than the Andromeda galaxy. It was fairly average.

In 1952, Baade's former Ph.D. student Allan Sandage joined the full-time staff at the Hale Observatories. Sandage, born in Oxford, Ohio, had known by age eleven that he wanted to be an astronomer when he grew up. When he looked through a friend's telescope, as he remembers it, "a firestorm took place in my brain." He spent two years at Miami University in Ohio and then two in the navy as an electronics specialist. When World War II ended, Sandage finished his undergraduate degree at the University of Illinois, then headed west and enrolled in 1948 as one of Cal Tech's first Ph.D. candidates in astronomy. While he was still working on his degree, he began collecting data at Mount Wilson for Baade, who was his thesis adviser. When Hubble, who was observing with the 200-inch, suffered a heart attack and needed a graduate student assistant to help him continue his work, Sandage was summoned to Mount Palomar. After a short sojourn at Princeton as a postdoctoral student, he came back as a member of the staff.

In 1958, five years after Hubble's death, Sandage discovered that some of what Hubble had thought were bright stars in distant galaxies and had used as measuring rods were instead glowing nebulae lit by many stars. That discovery more than tripled the size of the universe and increased its estimated age to about 13 billion years.

Human beings have a long history of underestimating the distance to objects in the heavens and the size of the universe as a whole. Though there have been a few overestimates, the scientific and popular image of the universe through the years has had to be revised upward, by startling increments, again and again. We have no intuitive feel for a distance of 13 billion light-years, the size Sandage settled on in the late 1950s. But was that still too small? Or had he actually taken things too far? Could it be that no one would ever know the answer? Nearly everyone who read the newspaper had been aware since the early 1930s that the universe wasn't cooperating as much as one might wish in this measuring venture. In more than one sense, it wasn't holding still to have its measurements taken!

CHAPTER

Coming Apart
in All Directions

{ 1 9 2 9 – 9 2 }

66 On philosophical grounds too I cannot see any good reason
for preferring the Big Bang idea. Indeed it seems to me in the
philosophical sense to be a distinctly unsatisfactory notion,
since it puts the basic assumption out of sight where it can
never be challenged by a direct appeal to observation. 99

Fred Hoyle

In the late 1920s, a sea change was about to occur in the way men
and women envisioned the universe. All of Edwin Hubble's earlier
contributions might seem enough for one lifetime, but when it came
to the impact he was to have on the history of human thought,
Hubble had only barely begun.

Hubble was continuing his observational work in collaboration
with Milton Humason, whose background, like Hubble's, was not in
science. Humason was a janitor and mule driver at Mount Wilson
whose formal education had ended at age fourteen when he'd come to
summer camp there and decided he didn't want to leave. By 1919, Humason, now twenty-eight, was a night assistant, tending the telescopes,

assisting the astronomers, and occasionally doing a little observing on his own. So obvious was his extraordinary innate skill with delicate instruments and the large telescopes that George Ellery Hale, ignoring the opposition of those who thought someone of Humason's background should not be part of the academic staff, decided to appoint him assistant astronomer. In 1928, Humason began working with Hubble on the measurement of redshifts of faint distant galaxies.

In 1929, having established beyond doubt that there are many galaxies besides our own, though Hubble was still not calling them "galaxies," the two men made the first announcement since Copernicus's *De Revolutionibus* that rivaled that book's significance in the history of astronomy. Hubble and Humason had found that except for galaxies clustered close to the Milky Way, every galaxy in the universe appears to be receding from Earth. What's more, on the large scale, every galaxy appears to be receding from every other. These discoveries had a much more immediate impact on the scientific community and the wider public than Copernicus's book had had, precipitating rapid change in ideas about what the universe is like, about its history, and even about ourselves.

It hadn't escaped Hubble's notice that there were connections between the observations that Slipher, Humason, and he were making and the solutions that physicists such as Willem de Sitter, Alexander Friedmann, and Abbé Georges-Henri Lemaître were getting from the equations of Albert Einstein—solutions that implied that the universe must be either expanding or contracting.

The Long-Delayed Demise of Constancy and Stability

The debate about whether the universe is expanding, shrinking, or just holding its own has a history that goes back long before the

Albert Einstein and Edwin Hubble during Einstein's visit to Mount Wilson in 1930. *(Huntington Library, San Marino, California)*

twentieth century. Ancient and medieval thinkers regarded the Earth as the region of the universe where change, decay, and evil held sway, while all beyond the Moon was unchanging and perfect. Newton echoed these sentiments to the extent of believing that a universe created by God could not be changing dramatically over time, for constancy and stability reflected the nature of God, whereas change (implying decay and conflict) did not. Newton thought that if a system goes far awry, as he realized the planetary orbits would over time, God would set it right again, so things would never be allowed to change too drastically.

Newton also argued on grounds of logic that the universe couldn't be expanding or contracting. He reasoned that if it were doing either, there would have to be a center to the motion—in other words, a point away from which it's expanding or toward which it's contracting. But matter distributed uniformly through an *infinite* space (as he believed it was) has no center. Newton couldn't foresee that others nearly three centuries later would find that his own equations lead to the prediction that the universe must be expanding or contracting. In the eighteenth century, Kant, who took off from Thomas Wright's picture of the universe as a flattened slab of stars, thought that if the universe were not perfectly balanced between the orbital motion of stars and their gravitational attraction for each other, it would end in destruction and chaos and lack "the character of that stability which is the mark of the choice of God."

Although it was less in fashion in the mid- to late 1800s and the 1900s to include God in scientific statements, the feeling that there was something sublimely rational and sacred about an unchanging universe and something shifty and distasteful about one that changed had by no means disappeared. It had become a doctrine of science rather than of religion. In an interesting turnabout, one of the twentieth-century reasons for clinging doggedly to the notion of a static universe was that an expanding universe—which almost surely must have had a beginning—seemed more likely to require a creator. That was a possibility some thought had been safely put to rest.

Albert Einstein resisted the idea of an expanding or contracting universe for reasons having to do with his scientific intuition. Soon after Einstein produced his general theory of relativity in 1915, he and de Sitter realized that solutions to Einstein's equations implied that the universe is either expanding or contracting. Einstein was not necessarily one to cling to old assumptions, but at this juncture he did, and he dug in his heels. Annoyed by the ridiculous upshot of his equations, he wrote, "To admit such a possibility seems senseless." Such strong aversion did he feel that he decided to adjust his theory to cancel out the offensive prediction. He put in a new constant of nature—a "cosmological constant," a mathematical term that would allow the universe to be static (not expanding or contracting). Later he regretted this move, calling it "the biggest blunder of my life." But the notion of a cosmological constant didn't disappear when Einstein reneged on it. It still haunts physics.

While Einstein was tinkering with his equations, Russian Alexander Friedmann decided instead to take Einstein's theory at face value. Friedmann insisted that if there is a cosmological constant, its value is probably nothing else but o (zero). He pointed out in one of his first papers dealing with Einstein's theories that the assumption that the universe is static had always been only an assumption. No observations required one to believe it. Einstein himself was well aware this was the case.

Friedmann proceeded to find not one, but a number of solutions to the cosmological equations of general relativity. Each solution described a different sort of universe. (See figure 6.1.)

Friedmann predicted that regardless of where you were to situate yourself in the universe, in any galaxy, you would find the other galaxies receding from you. The farther away a galaxy is from you, the faster it's receding, twice as far away, twice as fast. For an analogy, imagine a loaf of raisin bread rising in the oven. While the dough rises and expands between the raisins, every raisin sees every other raisin moving away from it, twice as far, twice as fast. No human being has been able to observe the universe from any vantage point except this solar system, but at least from here that is the sort of recession that Hubble observed in 1929 with the 100-inch telescope at Mount Wilson. The outward-bound speed of a galaxy is directly proportional to its distance from Earth. Twice as far away, twice as fast.

Belgian astrophysicist and theologian Abbé Georges-Henri Lemaître discovered solutions to Einstein's equations that were similar to Friedmann's. What intrigued Lemaître most was what the equations and solutions could reveal about the origin of the universe. It was Lemaître who first described something like what was soon to be dubbed, derogatorily, the "Big Bang," though he didn't give it that name. His suggestion was that there must have been a time when everything that makes up the present universe was compressed into a space only about thirty times the size of the

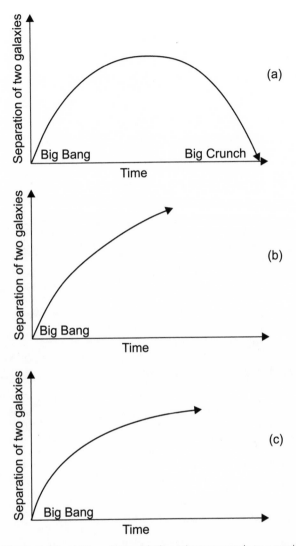

Figure 6.1 Three models of the universe: (a) The universe expands to a maximum size and then recollapses. (b) The universe expands rapidly and never stops expanding. (c) The universe expands at exactly a critical rate to avoid recollapse.

Sun—a "primeval atom." Partly because he was a priest and theologian as well as an astrophysicist, some of Lemaître's colleagues greeted his idea with derision. It smacked too much of Genesis.

Though Friedmann's theoretical work ended prematurely—he died at the age of thirty-seven—and he remained largely unknown except among mathematicians, Lemaître's work came to the attention of observational astronomers, largely through British physics giant Arthur Eddington, whose student Lemaître had been at Cambridge, and another of Eddington's students, George McVittie.

Thinking about the universe as an expanding lump of raisin bread dough led to interesting speculation. Would it be possible, assuming the necessary technology existed, to travel to the surface of the loaf and find the border of the universe? What would be beyond that? Unfortunately, those questions probably have no real meaning. Eddington fielded them by providing an analogy of a balloon and an ant:

The balloon has dots painted all over it. The ant crawls on the surface of the balloon. All that exists for this ant is that surface. It can't look outward from the balloon's surface or conceive of an interior to the balloon. Air is let into the balloon and the balloon expands. The ant sees every dot on the surface of the balloon moving away. Anywhere the ant crawls on the balloon, every dot is moving away. The ant may wander forever, like the Flying Dutchman, but it will never find an edge or a border to this universe. Our situation in our own universe is probably similar to the ant's, but with more dimensions. There is no edge from which we would see galaxies in one direction and absolutely nothing in the other.

What about the question of a "center"—the objection that if the universe were expanding or contracting there would have to be a center to the motion? Can anyone show where in the universe the expansion began . . . from what center point everything is retreating? The Big Bang was an explosion that sent everything flying outward.

Even granting that there are no absolute directions in the universe, it would seem that beings riding on a piece of debris from this explosion would have the right to assume there is an answer to the question: Where did the explosion take place in relation to where we are now?

Eddington's balloon analogy helps with that question as well. Can the ant ask where on the balloon's surface the expansion began? No. Watching the ant, you can see that that question would be meaningless. No dot on the balloon represents the "center" of the expansion. Newton failed to imagine a situation in which all points in the universe are moving away from all other points, with no "center" to the expansion, no direction in space toward which humans can look and insist that it all began there.

Paradoxically, living in an expanding universe means that there is a direction in which you *can* peer and see something different, perhaps even see an "edge." That direction is the past. What's more, in any space direction you look, you look toward the origin of the universe, for *any* direction is toward the past. Not only is that true when astronomers gaze deep into space with telescopes. It's true even in the small area of the room in which you read this paragraph. What you see of the opposite wall is old news. Of course the delay with which the picture of that wall reaches your eyes is not worth considering because light, and thus any picture that comes into your eyes, travels extremely fast, 186,000 miles or 300,000 kilometers per second.

When speaking of cosmic distances—where measurement in light-years is more meaningful than measurement in miles or kilometers—light speed *isn't* terribly fast, and the delay *is* worth considering. In fact, it can't be ignored. As the history of cosmic measurement continued in the twentieth century, measurement of distance in space would become inextricably bound up with measurement of distance in time. One can no longer ask how far away something is without also implying the other question: How far in the past am I seeing it? Questions about what is meant by an "edge" or "outside the universe" become entangled with questions about what is meant by a "beginning" or "before the universe."

In the 1930s, many astronomers and theoretical physicists were taking Hubble's observations as direct evidence that the universe is expanding, but there still was resistance to the idea and it was not all from within scientific circles. As was the case with Newton's *Principia*, the public was aware of stupendous changes going on in science. Einstein's theories were popularized in many forms, and his name became a household word. When a new discovery or theory is fundamental enough to impinge on everyone's concept of the universe and reality—not just a few specialized scientists—there tends to be a feeling that others besides scientists should have a say about what is True in this matter.

In 1936, in the Soviet Union, Joseph Stalin began a purge of scientists whose scientific findings and conclusions were not politically correct. One of the forbidden ideas was that the universe is expanding. In his book *Fireside Astronomy*, British astronomer Patrick Moore tells of the experiences of his friend Nikolai Kozyrev. Kozyrev was an astrophysicist at the Pulkovo Observatory near St. Petersburg. In November 1936, he was arrested and physically assaulted. In May 1937, he came to trial. The nature of his offense was never clearly stated, but he was sent to prison. After two years, Kozyrev ended up in a labor camp, and there a fellow prisoner reported him for holding scientific views about an expanding universe that were contrary to Soviet doctrine.

Kozyrev was resentenced to ten years' imprisonment. When he appealed, the sentence was changed to death. There was no firing squad at the labor camp, and a second appeal got the sentence reduced again to ten years. Gregory Shain, later the director of the Crimean Observatory, rescued Kozyrev from this appalling situation. He managed to get Kozyrev transferred to Moscow in 1945 and saw him set free in 1947. Kozyrev returned to his work in astrophysics, having lost ten years. Other Soviet scientists were less fortunate. Many were executed. The persecution even extended to the scientists' families. Kozyrev's wife was imprisoned, though not for such a long period as her husband.

Elsewhere the opposition was less extreme, and it soon focused not so much on whether the universe is expanding as upon whether it had a beginning. It was here that discoveries in astronomy and physics theory tread most seriously on philosophical and religious sensibilities.

One interpretation of Galileo's trial sees it as a clear contest between the authority of religion and the authority of science. In the twentieth century, those who had thought science had won that contest long ago were chagrined to find science suddenly seeming to uphold a religious point of view. Anyone for whom the idea of a God was anathema now had to face the unthinkable: a beginning . . . a moment of choice about whether there would be a universe . . . a creator.

Not that all who found the Big Bang philosophically disquieting were self-declared atheists. Many men and women, often without giving much thought to whether this conflicted with their religious beliefs, had put their trust in the power of science to explain the world. Since the time of Newton, it had been a growing assumption both in and out of science that scientific laws and explanations underlie everything that occurs, even those things that remain most mysterious and hidden, and that, given time, human minds ought to be able to discover those laws and explanations. The Big Bang threatened that cherished assumption.

A passage from Robert Jastrow's 1978 book *God and the Astronomers* sums up the situation. Jastrow is himself an astronomer and an agnostic, but he chides his colleagues for their reaction: "the re-

sponse of the scientific mind—supposedly a very objective mind—
when evidence uncovered by science itself leads to a conflict with
the articles of faith in our profession." He goes on to say:

> This is an exceedingly strange development, unexpected by all but the
> theologians. They have always accepted the word of the Bible: In the
> beginning God created heaven and earth. To which St. Augustine added,
> "who can understand this mystery or explain it to others?" The develop-
> ment is unexpected because science has had such extraordinary success
> in tracing the chain of cause and effect backward in time. . . . Now we
> would like to pursue that inquiry farther back in time, but the barrier to
> further progress seems insurmountable. It is not a matter of another
> year, another decade of work, another measurement, or another theory;
> at this moment it seems as though science will never be able to raise the
> curtain on the mystery of creation. For the scientist who has lived by his
> faith in the power of reason, the story ends like a bad dream. He has
> scaled the mountains of ignorance, he is about to conquer the highest
> peak; as he pulls himself over the final rock, he is greeted by a band of
> theologians who have been sitting there for centuries.

Three Cambridge University physicists declined the invitation
to sit down with the theologians. Just as there had been an alternative way of explaining Galileo's findings without having to have a
moving Earth—Tycho Brahe's model—there was an alternative way
of explaining Hubble's and Einstein's without having to have a beginning.

In 1948, Hermann Bondi and Thomas Gold, both originally
from Austria, and Fred Hoyle introduced theories that allowed the
expansion of the universe to happen without requiring that the universe have a beginning in time. Their Steady State theory became
the Big Bang's major competitor. According to Bondi, Gold, and
Hoyle's proposal, the universe hasn't always contained all the matter
that is in it today. As the universe expands, new matter emerges to

fill in the broadening gaps, and the average density of matter in the universe remains the same. While the stars in a galaxy like the Milky Way burn out and the galaxy dies, new galaxies are forming from the new matter. There would be no beginning or end to a Steady State universe. The unwelcome hint of "creation" suggested by Big Bang theory would be eradicated.

For at least two decades, the scientific and philosophical debate went on between those who favored one theory and those who insisted on the other, until finally, in the mid-1960s, new evidence came to light that Steady State theory could not explain and Big Bang theory actually had predicted. This evidence didn't come from an optical telescope. By then astronomers had discovered that the old phrase "I won't believe it until I see it" represented a ridiculously limiting attitude. Most of what goes on in the universe can't be "seen" at all. It happens beyond the visible range of the spectrum.

Beyond the Rainbow

It was no coincidence that studies of the heavens were for centuries limited to studies of visible light, and that radio astronomy was the first new astronomy to emerge. Radiation in those two parts of the electromagnetic spectrum can pass through the Earth's atmosphere. (Radio astronomy includes the microwave range.) Radiation in the infrared range can reach as low as the highest mountains and, at a few wavelengths, somewhat lower. But the Earth's atmosphere effectively blocks other radiation. Ultraviolet rays, X rays, and gamma rays coming from space have to be studied with telescopes on satellites above the atmosphere, and that wasn't possible until the late 1950s.

Radio astronomy began a quarter of a century before that, almost by accident. In the 1930s, trans-Atlantic phone calls took place

Karl Jansky and his rotating antenna. *(Bell Laboratories /Lucent Technologies Archives)*

by radio transmission and were plagued by static. The task of finding out what was causing the static fell to Karl Jansky of the Bell
Telephone Laboratories in Holmdel, New Jersey. He built a special
radio antenna—a long array of metal pipes—to aid him in his investigation. As Jansky sorted out the static, he found that most of it
came from thunderstorms, but there was also a faint hissing static
that couldn't be so easily explained. The hiss was strongest when
the region of the sky in the direction of the constellation Sagittarius
was overhead. Most astronomers agreed that the central regions of
the galaxy lie in that direction. When this part of the sky was below
the horizon, the hiss was weaker, but it never disappeared entirely.
The expectation prior to Jansky's discovery had been that the Sun
would be the strongest source of radio waves in the sky, just as it is
the brightest source of light. Now it seemed that a source could be
very "bright" in another part of the spectrum but show up not at all
in terms of visible light. Jansky concluded that he had discovered

the center of the galaxy. What else might be out there that optical telescopes were missing?

Though it might seem that such a discovery as Jansky's would have made headlines, his work received little attention even within the astronomy community. The one exception to this apathy was Grote Reber, an eccentric ham radio operator in Wheaton, Illinois, who read about Jansky's radio hiss in *Popular Astronomy* magazine. Reber proceeded to cash in his savings and design and construct his own device in his mother's backyard to listen to radio signals coming from the sky. Reber's telescope was a dish thirty feet (nine meters) in diameter. Astronomer Jesse Greenstein of Yerkes Observatory, when he later saw the backyard telescope, called Reber "the ideal American inventor. If he had not been interested in radio astronomy, he would have made a million dollars."

In the 1940s, Greenstein tried to get Reber a position at the University of Chicago. That fell through when the university agreed to have Reber only if his pay and research support came from Washington, and Reber refused to go to the bother of explaining to bureaucrats precisely how the money for new telescopes would be used.

Except for Reber, until the 1950s there was virtually no interest in radio astronomy in the United States. However, astronomers elsewhere in the world were quicker, after the development of radar during World War II, to appreciate the potential advantages of studying radiation in a part of the electromagnetic spectrum outside the visual range. In 1946, a radar signal was bounced off the Moon, and in the years that followed, the Jodrell Bank radio telescope in England led the world in mapping radio sources in space. Following close behind were Cambridge University and a team in Australia. Particularly intense sources of radiation in the radio range of the spectrum became known as "radio stars" and "radio galaxies."

A problem in early radio astronomy—and one reason it drew

little interest at first from optical astronomers—was that radio telescopes like Reber's couldn't measure a source's position in the sky
accurately enough to match up sources of radio waves with visible
objects. In order to accomplish that, there needed to be a hundred-
fold improvement in resolution, and that meant a telescope about a
kilometer in diameter. Radio waves come in a large range of lengths,
but they are all longer than waves in the visible part of the spectrum,
and their length is the problem. A radio telescope doesn't give good
resolution unless it is quite a bit larger than the length of the radio
waves it's receiving. Studying shorter radio waves only helps a little.
(By contrast, the much shorter waves in the visible part of the spectrum allow optical telescopes to achieve resolution with relative
ease.) Radio astronomers finally solved this problem in 1949 by
using networks of small telescopes linked to a receiving station that
combines the signals. Such a network or array is a "radio interferometer."

England and Australia continued to dominate the field. In 1950,
Martin Ryle of Cambridge University and his coworkers discovered
evidence of radio emissions from four nearby galaxies, including Andromeda. Ryle was an advocate of Big Bang theory, and as he and
his colleagues continued to map the distribution of radio sources
across the sky, their findings supported that theory. They discovered
that radio galaxies are much more abundant in the far distance than
they are nearby. Since the farther you peer into space the farther
back in time you are looking, Ryle's radio telescopes were showing
him the universe at a much earlier stage, and his discovery indicated
that the density of radio galaxies was greater then than it is today.
It's reasonable to think this should be the case if this is an expanding
universe in which everything on the large scale is moving away from
everything else and used to be much more closely crowded together.

In 1954, largely owing to the influence of Greenstein, by then a
professor at the California Institute of Technology, the National

Robert Wilson and Arno Penzias with the horn antenna at Bell Telephone Laboratories, Holmdel, New Jersey. *(Bell Laboratories/Lucent Technologies Archives)*

Radio Astronomy Observatory was built in West Virginia, and a radio interferometer was constructed near Yosemite National Park in California under the aegis of Cal Tech. But before that, even during the years when radio astronomy was virtually nonexistent in the United States, the Bell Telephone Company had continued to carry on research having to do with radio communications. It was at Bell Labs in Holmdel, New Jersey, where Jansky had first investigated the radio hiss from space in the 1930s, that Arno Penzias and Robert Wilson made the discovery that most scientists considered the clincher for the Big Bang theory.

Penzias was born to a Jewish family in Munich. When he was

six years old, he and his brother and parents were among the last Jews to get out of Nazi Germany, arriving as impoverished immigrants in the United States in the winter of 1940. Penzias received his undergraduate physics degree from the City College of New York and went on to Columbia University for his Ph.D. In 1961, he took a job at Bell Labs, where two years later he was joined by Texan Robert Wilson. Wilson had at first favored the Steady State theory over the Big Bang. He was strongly impressed with Fred Hoyle, who had been a visiting professor at Cal Tech when Wilson was a graduate student there. In 1963, a year after completing his Ph.D., Wilson crossed the country to the East Coast to work at Bell Labs.

At Bell there was a large, horn-shaped antenna designed for use with the *Echo I* communications satellite. In the spring of 1964, Penzias and Wilson were using the antenna to study noise levels that were hampering the satellite's transmission. Scientists working with the antenna had to make adjustments and limit themselves to signals that were stronger than the "noise." It was an annoyance that was possible to ignore, but Penzias and Wilson chose not to. They noticed that the noise remained the same regardless of which direction they pointed the antenna. If the Earth's atmosphere itself was the source of the noise, an antenna pointed toward the horizon should pick up more noise, for it faces more of that atmosphere than an antenna pointed straight up. Penzias and Wilson concluded that the noise had to be coming either from beyond the Earth's atmosphere or from the antenna itself. Pigeons nesting in the antenna were evicted and their droppings cleared away. That didn't help.

Penzias and Wilson were unaware of theoretical predictions that had been made back in the late 1940s and also of current work going on in England, Russia, and a few miles away from them in Princeton, New Jersey. In the 1940s, Russian-born physicist George Gamow, who had defected to the West in 1933, and Americans

Ralph Alpher and Robert Herman had theorized about the early universe, running Friedmann's equations backward toward the event with which the universe began. Their prediction was that there ought to be leftover radiation surviving from about 1,000 years after the origin of the universe. At that time, if Big Bang theory had it right, the universe was very hot, but by now the temperature of the radiation should have cooled to about five degrees above absolute zero. The prediction wasn't tested, for such radiation would not be easy to observe.

The idea that this radiation might indeed exist, and questions about its temperature, were still around in the early 1960s. In 1964, while Wilson and Penzias were tidying the pigeon droppings, Fred Hoyle and his colleague Roger Taylor in England were attempting to calculate what the background temperature of the universe would be today if it began in a Big Bang. And in the Soviet Union, Yakov Borisovich Zel'dovich had concluded that given the abundances of hydrogen, helium, and deuterium observed today, the universe must have begun in a hot Big Bang and its background temperature currently must be a few degrees above absolute zero. Soviet researchers had written articles about current radio astronomy measurements and what these measurements implied about the background radiation, and they had even suggested that the most likely antenna in the world to be able to detect this radiation was the Bell Labs antenna in Holmdel.

Robert Dicke at Princeton was working along the same lines. Born in Missouri in 1916, he was of Hoyle's generation. Dicke was educated at Princeton and at Rochester University, then worked on radar at MIT during World War II, and after that joined the Princeton faculty. On and off through the years, Dicke had considered the problems of the background temperature of the universe and the as-yet-undetected background radiation. In 1964, he set P. J. E. Peebles, a young researcher at Princeton, to work figuring out how the temperature might have changed over time in an expanding universe that had originated with a hot Big Bang. When Peebles had finished his calculations, Dicke gave two other researchers the job of setting up an antenna on the roof of the Princeton physics lab to try to detect the predicted radiation. It was at this juncture that Dicke received a phone call from Penzias and Wilson.

It so happened that Bernard Burke, another radio astronomer, had heard about Penzias and Wilson's puzzle, and he also knew (as they did not) of the work being done by Dicke. Burke proceeded to bring Dicke, Penzias, and Wilson together. They soon concluded that Penzias and Wilson had found by accident the radiation that Dicke had been hoping to discover.

This all-pervasive hum of radiation, coming at equal intensity from all over the sky, has been likened to a faded photograph of the universe as it existed about 300,000 years after the Big Bang. It is the oldest "photograph" we have, and the most direct evidence that the universe was once much hotter and denser than it is now. The radiation has cooled with the expansion of the universe and been redshifted so greatly that it reaches us in the microwave range of the spectrum at a temperature of about three degrees above absolute zero, a little cooler than the five degrees Gamow, Alpher, and Herman predicted in the 1940s.

What is the source of the radiation? According to Big Bang theory, the universe in its early stages was everywhere filled with electromagnetic radiation. This radiation wouldn't have been in the visible part of the spectrum. It was far too hot for that, in the trillions of degrees. As space expanded, stretching the wavelengths of the radiation, it shifted through the spectrum. The universe gradually cooled, but until the universe was about 300,000 years old the radiation was still too energetic to allow electrons and protons to bind together and form atoms. If an electron began to orbit a proton, it was knocked out of orbit by a photon—a particle of electromagnetic radiation. At about the 300,000-year mark, everything had cooled off enough so that photons no longer had the energy to knock electrons away from protons. Electrons and protons could form hydrogen nuclei and atoms, and photons could move about more freely. Physicists call this the "decoupling" of radiation and matter. With that change, radiation (photons) streamed in all directions, and it is that radiation, redshifted all the way across the spectrum to the microwave range, that we detect today as the cosmic microwave background radiation.

Though that radiation began its journey longer ago and farther away than anything else modern astronomers observe, you don't need special equipment to detect it. The snow on a TV screen that appears when a station isn't broadcasting consists in part of this radiation, this dim afterglow of the Big Bang cataclysm.

It's an interesting bit of astronomy trivia that Wilson and Penzias were not actually the first to detect and measure the cosmic microwave background radiation. In 1961, another engineer at Bell Labs, Ed Ohm, had also tried to figure out what was causing the "noise." Eliminating everything that could be explained away, he found that it was the equivalent of radiation at a temperature of about three degrees above absolute zero. Unfortunately for Ohm, he didn't find the problem as annoying as Wilson and Penzias did, and he didn't pursue it or suspect its significance. No one put him in touch with anyone who knew of the theoretical predictions. It was Wilson and Penzias who shared a Nobel Prize for the discovery.

Why wasn't this discovery made earlier? Gamow, Alpher, Herman, or Dicke probably could have discovered the cosmic microwave background radiation earlier had they tried. Ohm did discover it, but failed to realize what he'd found. The apparatus Dicke and his colleagues were constructing at Princeton would have detected the radiation and was later used to study it. But as it turned out,

the discovery involved the combination of a private corporation—
American Telephone and Telegraph (AT&T)—farsighted enough
to fund less practical science and attract such researchers as Wilson
and Penzias, the stubborn curiosity of these two men themselves—
who, unlike Ohm, weren't satisfied until they got to the bottom of
a mystery—and a serendipitous meeting of observation and theoretical understanding when those *almost* were ships passing in the
night.

The Big Bang Rakes in the Chips

From the early 1960s, everything seemed to fall into place for those
who favored Big Bang theory. More support came from the discovery that quasars—which theorists realized might very well be an
early stage of galaxy formation—exist only at enormous distances
from Earth. According to Steady State theory, galaxies are periodically dying and being replaced by new galaxies made from new matter. If that were so, and if quasars are part of the process of galaxy
formation, quasars ought to be fairly evenly distributed near and far
throughout the universe. The fact that they are not argues against
the Steady State and in favor of the Big Bang. Quasars' distance
from Earth in space (and, by virtue of that fact, in time) means they
existed only when the universe was much younger than it is now,
indicating that this particular stage of galaxy formation occurred
only in the distant past, hasn't happened again in later periods of
the universe's history, and isn't still going on today. The universe is
not repeating itself.

Astronomers and physicists have also continued to study the
cosmic microwave background radiation for clues about how the
universe has evolved. In 1973, Paul Richards and colleagues at Berkeley undertook balloon experiments to find out whether the spectrum

of the background radiation was the spectrum that Big Bang theory predicted. They found that it was.

Even more support for the theory, also in the early 1970s, came from studies of the spectra of other galaxies to measure the abundances of various elements in them. Big Bang theory had predicted that about 25 percent of the mass of all the elements making up the universe should be helium 4. The studies showed that prediction was on target. So did measurements within the Galaxy. Predictions of abundances of other elements such as deuterium, helium 3, and lithium also turned out to be what the theory prescribed.

While it looked increasingly likely that Steady State theory would go the way of Tycho Brahe's valiant last-ditch efforts to save the Earth-centered universe, the Big Bang theory wasn't entirely problem free either. Two stumbling blocks were the "horizon problem" and the "flatness problem."

The horizon problem stems from the observation that the cosmic microwave background radiation is very homogeneous, the same in all directions in areas of the universe too far separated for radiation ever to have passed from one to the other even at the earliest moments after the Big Bang. The intensity of radiation is so close to identical in those remote areas that it seems they must have exchanged energy and come to equilibrium. The question is: How?

The flatness problem has to do with why the universe has not either long ago collapsed again to a Big Crunch or else experienced such runaway expansion that gravity wouldn't have been able to pull any matter together to form stars. Neither has happened or seems to be happening at the moment. Yet having a universe somehow poised between those possibilities is so unlikely as to boggle the imagination. It would require the expansive energy (resulting from the Big Bang) and the force of gravity to be so close to equal that they would differ from equality by no more than 1 in 10^{60} (1 followed by sixty zeros) at a time less than 10^{-43} seconds after the Big

Bang (a fraction with 1 as the numerator and 1 with 43 zeros as the denominator).

A revised history of the Big Bang universe called "inflation theory" proposed to solve both those problems. The idea emerged in the late 1970s, when Alan Guth, then a young physicist at the Stanford Linear Accelerator, reached the conclusion that the universe might early on have undergone a period of stupendous growth before settling down to the expansion rate it has today. Guth knew immediately that he'd hit upon a good thing. "SPECTACULAR REALIZATION" he wrote in his notebook and drew two concentric boxes around the letters.

Guth proceeded to work out a process that, at a time less than 10^{-30} seconds after the Big Bang (a fraction with 1 as the numerator and 1 followed by thirty zeros as the denominator), could have caused gravity to become an enormous repulsive force. Instead of pulling matter back and slowing the expansion of the universe, it would, during a period lasting only an unimaginably small fraction of a second, have accelerated the expansion, causing violent, runaway inflation in the dimensions of the universe from a size smaller than a proton in the nucleus of an atom to about the size of a golf ball. When the inflationary period ended, the universe would continue to expand, but in the more sedate, familiar fashion.

Inflation theory helps the horizon problem by allowing the visible universe to have emerged from a region so tiny that it had the opportunity to reach equilibrium before it inflated, producing the homogeneity in the background radiation. When it comes to the flatness problem, inflation theory says that out of an infinite number of possible universe stories—those that have the universe collapsing, those that have it expanding forever to thin oblivion, and the one seemingly improbable story in which it is perfectly poised between the two—the most likely story is that the universe *will* be balanced between expansion and collapse, expanding forever but at a continually decreasing rate, neither collapsing nor eternally thinning out. Prior to inflation theory, Big Bang theory had not been able to help the universe walk that tightrope. But Guth and others who have contributed to the development of inflation theory explain that any imbalance between the expansive energy resulting from the Big Bang and the contracting force of gravity would have been wiped out by the period of runaway inflation, leaving the universe in that extremely unlikely and highly desirable condition of flatness. Highly desirable because that is the only sort of universe that eventually allows intelligent life to emerge.

To visualize the version of inflation theory that has the most to offer in this regard, first imagine the universe before the period of inflation begins, again using a balloon as an analogy. Inflate the balloon a little. That represents the expansion of the universe before the inflationary period. Pause to mark a tiny red dot on the surface of the balloon. Next attach the balloon to one of those machines that inflate balloons rapidly and turn the machine on at maximum force. That inflates the balloon to a truly remarkable size. The tiny red dot itself becomes huge. Imagine now that it isn't the whole balloon that represents everything that human beings observe and ever will observe of the universe; it is the red dot. Inflation theory asks us to believe that what we normally call "the universe" may be similarly only a tiny fraction of everything there is.

Imagine that instead of just one red dot, you have drawn dots all over the balloon, perhaps an infinite number of them. Having blown up the balloon to enormous dimensions, will you find every dot representing something equal in size to the observable universe?

Andrei Linde of Stanford University has suggested that may not be the case. Linde, a graduate of Moscow University and the P. N. Lebedev Physics Institute in Moscow, already had an enviable reputation in physics when he moved to Stanford in 1990. Among his many accomplishments—he also dabbles in sleight-of-hand magic, acrobatics, and hypnosis—Linde had introduced a new version of inflation theory in 1983. His proposal was that the early universe before the period of inflation was in a chaotic condition, something like the surface of the ocean. If the universe was this chaotic, it would be ridiculous to talk about *the* initial state of the universe. One could find all sorts of initial states, depending upon which bit of it was under examination. Start-up conditions for any of the dots on the balloon might be different from any other. The upshot is that when the gravitational repulsive force came, each dot would respond differently. Some not at all. But one version of inflation theory predicts that when the inflation ended, you would find that in any dot that *had* inflated, the force of gravity (now working in the more familiar way) and the repulsive force resulting from the original Big Bang explosion would be balanced in the way we experience today in our universe. Perhaps only one of all those dots would have been able to end up balanced in that way. If so, that dot is our universe.

Though inflation theory had great success explaining away problems in Big Bang theory, there was no observational evidence to assure anyone that inflation actually is *the correct* explanation. Discoveries in the late 1990s would throw theorists a new set of problems and perhaps new solutions, as well as some hope of observational evidence.

A third stubborn puzzle plaguing Big Bang theorists after Wilson and Penzias's discovery was how a universe that looked so uniform when it was 300,000 years old had become so diverse and clumpy all these years later. In repeated measurements, researchers found that the cosmic microwave background radiation was disappointingly uniform in temperature. The temperature was the same in readings taken out to the end of observability in every direction. This meant the early universe must have been extremely smooth, without lumps, clumps, or irregularities that would show up as fluctuations in that temperature. How, then, could the universe have evolved to have galaxy clusters, galaxies, stars, planets—even such small clumps of matter as people? Somewhere back there must lie the seeds of those developments, but where?

Here is the problem: Picture every particle of matter in the universe attracting every other by means of gravitational attraction. The closer to one another the particles are, the stronger they feel each other's gravitational pull. If all particles of matter in the universe are equidistant, and there are no areas in which a few particles have drawn together even slightly more densely, then every particle will feel equal pull from every direction and none will budge to move closer to any other particle. It was this sort of gridlock researchers seemed to have discovered in the early universe, where matter appeared to have been distributed so evenly that it could never yield to form the structure evident in the universe today. If this were not so, why couldn't anyone find even the tiniest fluctuation in the background radiation—the "photograph" of how matter was distributed back then?

Finally, in April 1992, astrophysicist George Smoot at Lawrence Berkeley Laboratory and the University of California at Berkeley announced that he and his cohorts at several other institutions had found the long-sought "wrinkles." New data from a satellite called the Cosmic Background Explorer (COBE) had revealed the fluctuations in the cosmic microwave background radiation that astrophysicists had been seeking for a quarter of a century. The temperature fluctuations measured no more than a hundred-thousandth of a degree, but that was enough, the researchers felt, to explain what had happened to the universe. These minuscule variations in its topography when it was only 300,000 years old were evidence of a gravitational situation in which matter could have attracted matter into larger and larger clumps.

How did the tiny fluctuations get there? There have been some stabs at answering the question, but no observational evidence to help. Inflation theorists point out that according to the so-called

Heisenberg uncertainty principle of quantum mechanics, what we call "empty space" cannot actually be empty. Instead, always and everywhere in the universe there are tiny energy fluctuations. During the inflation period, the peaks and troughs generated by these fluctuations in the newborn universe would have been inflated enough to serve as the seeds of all the irregularity to come.

In the year 2000, NASA is scheduled to launch its Microwave Anisotropy Probe. "MAP" will measure the background radiation thirty times more precisely than the Cosmic Background Explorer did. Even more precise observations should come from the European Space Agency's Planck Satellite beginning in 2004. There are also several balloon-based missions in the offing.

Meanwhile, a wealth of evidence points to the fact that we do live in a Big Bang universe. To sum up its history, according to the theory as it stands amended by inflation theory: All that human beings observe or ever will be able to observe started out compressed in a state of almost unimaginable density. That exploded and everything—space itself—began to expand. After a short interval of extremely rapid expansion, the expansion slowed down and continued more sedately. Everything thinned out and cooled. All was smooth and virtually uniform except for some faint wrinkles, or "density fluctuations." While expansion continued, the gravitational attraction of areas where matter was already concentrated more densely pulled in more matter, and thus matter began clumping, eventually forming stars, galaxies, clusters, and superclusters of galaxies gravitationally bound to one another.

Confronting a Gordian Knot

While observations and experiments were confirming earlier Big Bang predictions, theorists had been raising new questions about the moment of "beginning" itself. One in particular was: Does an expanding universe that is not a Steady State universe have to have had a beginning?

Both everyday logic and mathematics seem to indicate that in a universe where on a large scale everything is moving farther and farther from everything else, if you could reverse the direction of time and travel back toward the beginning, you would find things getting closer and closer together. Eventually everything would be in precisely the same place. Are there any other possibilities?

In 1963, Russian scientists Evgenii Lifshitz and Isaac Khalatnikov had proposed another outcome to this time-reversed story. They ran the history-of-the-universe film backward, imagining a scenario in which the universe contracts and all the galaxies draw closer to one another, appearing to be on a collision course. But Lifshitz and Khalatnikov pointed out that the galaxies have other motion in addition to the motion that brings them toward one another. Could it be that this additional motion might cause them, as they approach one another, to miss one another and fly past with a nod, so to speak? Was this the way to avoid having a beginning? If you kept watching the movie backward, might you simply see the universe expand again? Or must everything have begun in the same spot?

It was this question that engaged Stephen Hawking of Cambridge and Roger Penrose of Oxford in the middle and late 1960s.

Hawking's story is well known. He has become a legend in his time—the physicist who explores the frontiers of scientific knowledge and speculation, who works by spectacular leaps of intuition, who often seems to reverse himself, who writes best-selling books that readers struggle to understand. Lou Gehrig's disease has locked Hawking's body motionless in a wheelchair, but he is the most nimble-minded of men.

Roger Penrose also thinks outside the envelope. In his youth he discovered an "impossible object"—which means a figure that can't really exist because it contradicts itself. His father helped him turn the idea into the "Penrose Staircase," and Maurits Escher used it in two of his famous lithographs: *Ascending and Descending*, and *Waterfall*. Penrose also managed to visualize an impossible object in four-dimensional space. As he matured, he never gave up "playful" mathematics. He's become one of the world's most imaginative mathematicians, physicists, and authors, and has discovered two shapes ("Penrose Tiles") that in their three-dimensional forms may underlie a new kind of matter.

In 1965, Penrose, building on earlier work of John Archibald Wheeler, Subrahmanyan Chandrasekhar, and others, showed that if the universe obeys general relativity and several other constraints, when a very massive star has no nuclear fuel left to burn and collapses under the force of its own gravity, it will be crushed to a point of infinite density and infinite space-time curvature—a "singularity." General relativity had predicted the existence of singularities, and in the early 1960s physicists had speculated that a star of great enough mass undergoing gravitational collapse might form a singularity at the center of a black hole, but very few had taken this prediction seriously. Penrose calculated that this will happen even if the collapse isn't perfectly smooth and symmetrical. No "might" about it. It must.

Hawking, in his doctoral thesis at Cambridge in 1965, reversed the direction of time and applied the same concept to the entire universe, suspecting that if he could watch the expansion of the universe run backward he would discover something similar to what Penrose had found with black holes. Once the collapse (the expansion of the universe run in reverse) had proceeded far enough, whatever additional motions the galaxies had would make no difference to the history of the universe. By 1970, Hawking and Penrose were able to demonstrate, in Hawking's words, "that if general relativity is correct, any reasonable model of the universe must start with a singularity." Everything that was to be the matter/energy of the universe that human beings might eventually be able to observe would have been compressed not to the sphere Lemaître envisioned (the primeval atom) but to something much smaller than that—to a point of infinite density.

Well, that did it! Physical theories can't work with infinite numbers. When the theory of general relativity predicts a singularity of infinite density and infinite space-time curvature, it also predicts its own breakdown. All the theories of classical physics are useless at a singularity. There is no possibility of predicting what will emerge; one can only wait to observe what it will be. Indeed, why should anything emerge at all? There is no way to find out why this singularity suddenly ceases to be a singularity and becomes a universe. And what happened before the singularity? It's not even clear whether that question has any meaning.

Hawking and Penrose's discovery did not, however, put an end to attempts to devise an origin-of-the-universe story that would be more palatable to those unwilling to accept unanswerable questions and hints of a creator. Hawking would soon be one of those most eagerly trying to untie the Gordian Knot he and Penrose had discovered.

Deciphering
Ancient Light

{1946–99}

GG When people on airplanes ask me what I do, I used to say I was a physicist, which ended the discussion. I once said I was a cosmologist, but they started asking me about makeup, and the title 'astronomer' gets confused with astrologer. Now I say I make maps. ⟩⟩

Margaret Geller

In the second half of the twentieth century, the cosmic distance ladder became increasingly extensive and sturdy. Measurements using old and new techniques served as checks on one another, and previous estimates were reexamined with improved technology and fresh theoretical understanding. It became possible to discover more than ever before about the structure of the Galaxy and of the universe beyond.

In 1946, researchers had first bounced a radar signal off the Moon. After that, astronomers were able to fine-tune distance measurements within the solar system by bouncing signals off the planets and the outer atmosphere of the Sun and timing how long it took the radar echos to come back. Later, unmanned missions visited the planets. If you send a spacecraft to Mars, and it gets there with only minor course corrections along the way, you know you have Mars's distance and orbit more or less right. That settles any argument between you and Cassini.

Between the late 1830s—when Bessel, Henderson, and Struve first measured stellar parallax—and 1900, astronomers used the parallax method to measure approximate distances to no more than about 100 stars. Because the Earth's atmosphere refracts light rays passing through, blurring the images of stars, even today the best ground-based telescopes measure accurate parallaxes of the brightest stars out to only about 300 light-years, a minuscule distance by the standards of the Galaxy. The potential for considerable extension of that range came in the late 1950s with the ability to put telescopes beyond the atmosphere. By the early 1990s, astronomers had reasonably accurate parallaxes for close to 10,000 stars. In 1989, the European Space Agency launched the satellite *Hipparcos* (High Precision Parallax Collecting Satellite). *Hipparcos* still measures parallaxes from the baseline provided by the Earth's orbit, but it can measure them several times farther away than earthbound telescopes, in a volume of space a hundred times as large. By the mid-1990s, *Hipparcos* had swelled the catalog of precisely measured distances to 120,000 stars. Before it came on-line, there was still no more accurate way to determine the distance to the nearest star cluster, the Hyades, than the old moving cluster method. *Hipparcos* measures its parallax directly.

Historically, one of the most important advances in cosmic measurement had been the discovery that Cepheid variables could serve as standard candles, distance calibrators to measure beyond the parallax range. However, when it came to finding absolute distances, this rung in the cosmic distance ladder depended on there

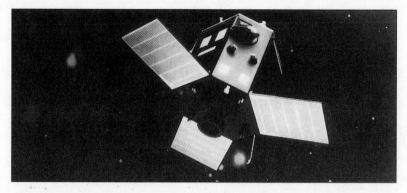

Hipparcos (High Precision Parallax Collecting Satellite). *(European Space Agency)*

being Cepheids close enough to be measured directly by parallax. There were none. The best that could be done was to measure their distance by a version of statistical parallax, which meant that the Cepheid rung was a less reliable part of the ladder than it might have been if there were closer Cepheids. Again the *Hipparcos* satellite finally promises a breakthrough. It has now measured a few Cepheid parallaxes directly. There is some question about the reliability of these measurements, but astronomers hope these uncertainties will be resolved by a project called the Space Interferometry Mission in the first decade of the twenty-first century.

Experts have studied the physics of stars, the rate at which they convert hydrogen to helium and burn up their fuel, their masses, their temperatures, their life cycles, their spectra, the way they affect one another in binary and ternary groupings, the idiosyncrasies that set some of them apart, and the wobbling that indicates there are planets orbiting a star. But none of this burgeoning knowledge has diminished the significance of Cepheids. They are still arguably the best hope for establishing absolute distance measurements far into the universe. The Cepheids Leavitt first studied in the Magellanic Clouds were about 169,000 light-years away. With modern ground-based telescopes, astronomers detect Cepheids in galaxies 15 million light-years distant. With the Hubble Space Telescope, NASA's orbiting observatory that came into full use after repairs in December 1993, researchers study them in galaxies almost 60 million light-years away. The Hubble Telescope's mirror is smaller than many ground-based telescopes, but because it's based outside the distorting atmosphere of the Earth, it concentrates the light from stars into images that are many times sharper.

Looking at stars' spectra has proved to be among the most effective ways to gain knowledge about them. In the past few decades, astronomers have continued to study the spectra of thousands of stars whose distances are known from various methods. Working from a discovery of Danish astronomer Ejnar Hertzsprung's that for stars of any particular spectral type there is a correlation between the width of the spectral lines and a star's absolute magnitude, experts have succeeded in compiling tables giving the absolute magnitude for stars with *any* combination of spectral type and line width. Check a star's spectral lines, look up the absolute magnitude it should have, compare this with its apparent magnitude, and you know its distance! The method is called "spectroscopic parallax." Its precise accuracy depends on how correct the measured distances were for the stars used to compile the table, but astronomers believe it is reliable. Not only does this allow one to measure the distance to any star for which a spectrum can be obtained, it also makes it possible to find the distance to a cloud of gas or dust by finding a star in it and measuring that star's distance.

Another tool that, like spectroscopy, was inherited from the nineteenth century and put to optimum use in the twentieth is the Doppler shift. When Slipher measured the redshifts of the nebulae in the 1920s, he used the twenty-four-inch telescope at the Lowell Observatory, which was one of the finest of that day. But Slipher had to spend night after night in the unheated telescope dome, exposing each photograph twenty to forty hours to obtain spectra from which he then could measure the shifts. The equipment used by Hubble was better, but he and his contemporaries still had to expose a single photographic plate to hours of light coming through a telescope to record the tiny spectrum that they could then study with a magnifier to find the spectral lines. Today, the job of finding the redshifts of galaxies and quasars is done in minutes by telescopes equipped with charge-coupled devices (CCDs)—silicon chips that convert light from the night sky into digitized images. Computer-run arrays make it possible to take the spectra of many galaxies at once.

Redshift has become the tool of choice, and often also the tool of necessity, for measuring the distant universe, and much of the mapping on the largest scales has been based on redshift measurements only. By midcentury, astronomers had measured redshifts for about 100 galaxies. By 1970, the number was around 2,000. Today, redshifts have been cataloged for more than 100,000 galaxies and counting. There are ongoing projects such as the Sloan Digital Sky Survey, which alone is expected to collect the redshifts of 1 million galaxies.

Redshift measurement is not, however, infallible. How much the light from an object is redshifted gives an indication of how rapidly the object is increasing its distance from Earth, providing a basis for comparing the distances of far-off objects. The method would work flawlessly in an ideal situation where everything was moving directly away from everything else with the expansion of the universe (like the raisins in the rising loaf of raisin bread) and there was no other motion going on. Unfortunately, the situation is not that simple.

Galaxies do more than recede from one another with the expansion of the universe. Some of them are locked in binary systems, with two galaxies orbiting a common center of mass, so that at any one time, one of the pair is likely to be moving away from Earth at a speed greater than the simple expansion of the universe and the other at a speed less than that expansion rate, though they are really about the same distance away. Spiral galaxies also spin like giant pinwheels, and if a binary system is made up of two spiral galaxies, they probably spin in opposite directions, causing yet more complication in the overall motion. Pairs of galaxies and single galaxies are often grouped together in clusters, orbiting the gravitational center of the cluster. Similarly, clusters may be grouped together in superclusters, and all of these—galaxies, clusters, and superclusters—are pulling and tugging at themselves and one another by means of their gravitational attraction. Since it's extremely difficult to take all of this motion into account and interpret it correctly, measurements derived from redshift are continually subject to fine-tuning.

Halton C. Arp, who has spent many years at Mount Wilson Observatory studying galaxies, has cast a minority vote on the question of how much redshift should be trusted as a way of measuring distance to far-off galaxies and quasars. With Geoffrey Burbidge of the University of California, Arp has studied the brightest quasar, 3C273. Its redshift indicates that it should be about 2 billion light-years away. Arp has found, however, that 3C273 appears to be interacting with a giant elliptical cloud of hydrogen gas no more than 65 million light-years away. There are other mysterious cases where objects at dramatically different redshifts appear to be linked. Arp believes that many quasars are not so remote as most astronomers have been measuring them. Most of Arp's colleagues pass off his evidence as coincidence.

While astronomy was probing deeper and deeper into space as the twentieth century progressed, at a rate that makes earlier centuries look sleepy by comparison, it was also meeting some frustrating obstacles. Within the Milky Way Galaxy, visibility with the naked eye and optical telescopes is obscured by blotchy clouds of interstellar dust, especially in the direction of the Galactic center. Jansky had discovered that those clouds don't block radio waves. Was there more to be observed at other wavelengths? The revelations of early radio astronomy led many people to suspect that there might be much more to the universe than any earth-bound telescope, optical or radio, would ever detect. World War II radar had helped usher in the age of radio astronomy. It took another impetus from the arena of international politics to boost telescopes above the Earth's atmosphere.

The *Sputnik* Legacy

Space-based astronomy began as a distant by-product of the Treaty of Versailles, which ended World War I. That treaty limited Germany to the production of artillery of small caliber. Germany turned its research efforts and funds to rocketry instead of artillery and made significant progress in this new area. One result was the V-2 rockets used to attack southern England in 1944 and 1945, during World War II. A captured stock of V-2 rockets ended up in the United States, and twenty-five of them were set aside for scientific use. Projects were soon under way to develop controls that could keep these rockets positioned accurately and stably enough for astronomical observations to be made from them.

By the late 1950s, Western technology had improved steadily, and scientists were becoming accustomed to large budgets justified by the Cold War arms race and national security interests, even for projects that had no immediately obvious practical applications. A partnership of unprecedented scope between government and big-budget science had begun. However, a surprise from the Soviet Union jump-started Western space-age astronomy in earnest. On October 4, 1957, the Soviet Union launched *Sputnik I*, the first human-made satellite to orbit the Earth. The perception was that the Soviet Union had pulled ahead in the race into space and, by implication, in the arms race as well. To catch up, the Western nations, particularly the United States, began to pour money into the development of technology for spacecraft and satellites, and also into science education. (I was a teenager then, and I must admit I can't recall there *being* a race into space until suddenly "they" were winning. The panic filtered down to the level of my high school in Texas which wasn't, evidently, as good at producing scientists as Soviet schools were. Better physics and math books and lab equipment were purchased, teachers went to workshops for retraining, and we were all urged to take more classes in these subjects. My younger brother's stock, as a fledgling physicist and computer whiz, went up. Mine as a classical musician went down. At the University of Texas, my later-to-be husband was called unpatriotic by his mathematics professor for choosing not to major in math.)

Before long the Western nations also had sent spacecraft into orbit, and the race continued, to the enormous benefit of all astronomy, not just space-based projects. Cold War competition paid the escalating bills. Much had changed since 1609, when Galileo put together his own perspicillum, and the nineteenth century, when Lord Rosse financed the construction of his "Leviathan" and a university community pooled their resources to buy Harvard's telescope. In the early twentieth century, a portion of Lowell's fortune was enough to build a major observatory. A corporation, AT&T, paid for the telescope at Bell Labs. But the cost of astronomy later in the century was on another scale entirely, which strained even the largest national budgets. Those nations with the wherewithal also continued to have motivation for supporting science, other than the furthering of human knowledge. The technology that allowed telescopes to fly above the atmosphere, pointing upward, also made it possible for surveillance cameras to point downward and facilitated vastly improved worldwide communications systems. The technology that built and guided space probes also built and guided missiles. Moreover, the world image of a modern superpower included, just as it had on a smaller scale for the Medicis in Renaissance Florence and Louis XIV of France, an image of technological and scientific preeminence.

When telescopes and cameras probed for the first time above the atmosphere in the late 1950s and early 1960s, those who put them there found there was a great deal to be seen. The full potential of space-based astronomy took years to come to fruition and is still not fully realized. Nevertheless, a new era had begun. It was possible at last to view the universe without the refraction problem that had frustrated astronomers for centuries, and to make observations at many different wavelengths, comparing findings in one range with those in another. There were indeed objects out there that don't radiate at all at optical or radio wavelengths and others that look very different when viewed in other parts of the spectrum.

Unfortunately for our understanding of the Milky Way Galaxy, even all of this improved vision did not provide an easy way to calculate the distances to hot nebulae and clouds of gas in which stars are born, and it didn't allow astronomers to see the Galaxy from the outside. Though they would in time find ways to compile maps that are almost as good as seeing the Galaxy from afar, it was still easier to study things beyond the Galaxy.

Outside the Milky Way

For a while, the most significant progress in research outside the Galaxy continued to be made not from space but with ground-based telescopes. Hubble had used galaxies themselves as standard candles.

Other astronomers had chosen to assume that the brightest galaxy in a major cluster has approximately the same absolute magnitude as the brightest galaxy in every other major cluster. As it turned out, this method had some pitfalls. If you see two candles at a distance from you at night, and know that the two would have identical brightness if you saw them close-up, you can judge their distance relative to one another by how bright they appear to you—unless without realizing it you happen to view one of the candles just as a moth has flown into its flame and caused it to flare unusually bright. There is a risk of something like that happening when researchers use the brightest galaxies in galaxy clusters as standard candles. In a crowded cluster, it isn't uncommon for galaxies' paths to cross or come near enough so that a large galaxy "cannibalizes" a smaller one, making the cannibal for a time much brighter than normal. The "temporary" change can last a few hundred million years.

Addressing this and other problems, Allan Sandage, along with Gustav Tammann from Switzerland, carried cosmic-distance-ladder construction to unprecedented heights. Beginning in the early 1960s, Sandage and Tammann began using the 200-inch Mount Palomar telescope to set in place carefully calculated rungs leading deeper and deeper into the universe. They first utilized Cepheids to estimate distances to galaxies in the Local Group, the group that includes the Milky Way, Andromeda, and other relatively nearby galaxies. Their next step, also with Cepheids, got them to a great spiral galaxy, NGC2403, in another "group" called M81. With these distances in hand, they proceeded to measure the size and luminosity of huge clouds of ionized hydrogen gas in these galaxies, and to study how the size and luminosity of these clouds are related to the overall luminosity of a galaxy. That relationship gave them a way to calculate the distances of more remote galaxies containing similar gaseous clouds.

Sandage and Tammann also picked up on a new scheme from

The 200-inch Hale Telescope
on Mount Palomar.
*(California Institute of
Technology)*

Allan Sandage with the 100-inch telescope at
Mount Wilson. *(Frederic Golden)*

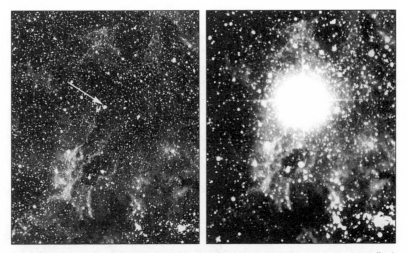

Supernova 1987A. Two photographs showing the same star field in the Large Magellanic Cloud, before and after the explosion. *(D. Malin, Anglo-Australian Telescope Board)*

Canadian Sidney van der Bergh for classifying spiral galaxies. Van der Bergh estimated their luminosity from the clarity and contrast of the spiral arms. Here was yet another standard candle to gauge distances of even more remote galaxies. By 1975, Sandage and Tammann had concluded that the universe was possibly as old as 18 billion years—a considerable increase over the 13 billion years Sandage had estimated in 1958 and a highly satisfactory result, because it allowed the universe to be sufficiently old to encompass the age of its oldest stars, even the most ancient globular clusters.

Sandage and others also found supernovae increasingly useful as yardsticks. Supernovae are exploding stars, and they are extremely bright, which makes them the easiest stars to observe at great distances. One estimate has it that in a one-minute interval, a supernova can put out more energy than all the "normal" stars in the observable universe. Only a portion of this energy is in the visible part of the spectrum, but even that can be sufficient to outshine the entire galaxy in which the supernova occurs. It goes without saying that such events could be remarkable standard candles—if they are in any way standard. If, for instance, all supernovae reached the same maximum brightness, that would provide an excellent way to find out how the distance to one supernova compares with the distance to another. Or it might be possible to understand individual explosions sufficiently well to estimate their distances even if each is unique. There has been work on both these fronts.

It turns out that all supernovae do not reach the same brightness. For a while astronomers thought one particular class—Type Ia—did. It was a setback in the 1990s to find there were brightness differences among Type Ia supernovae, but fortunately those differences soon became well enough understood to restore Type Ia's to their status as standard candles. Finding the distance to other types of supernovae is still problematic, but analysis of emissions in different wavelengths enables experts to find what type of star has exploded and its mass, as well as to study detailed properties of the aftermath of the explosion. The blast typically produces radiation at many wavelengths as it expands and meets surrounding matter.

Supernovae fall into two broad groups called Type I and Type II. Type I supernovae are exploding "white dwarf" stars. The story of such an event begins with an elderly star that has exhausted its nuclear fuel and collapsed to a sphere about the size of the Earth, with a mass probably close to or less than the mass of the Sun. The matter of which the star is composed is packed to almost inconceivable density.

Many white dwarf stars don't lead solitary lives but are part of binary systems in which two stars circle each other, orbiting their common center of mass. Often a dwarf star's partner is a large, far less dense star. As they circle, the gravitational pull of the denser dwarf star cannibalizes matter from its companion and the dwarf gradually puts on weight (mass). Eventually it tips the scales at 1.4 times the mass of the Sun. That mass is called the Chandrasekhar limit after Subrahmanyan Chandrasekhar. As a young theoretical physicist from India continuing his work in England, he calculated that limit in the early 1930s.

The white dwarf star, having no nuclear fuel to burn and having boosted its mass over the Chandrasekhar limit, collapses under the pull of its own gravity and rips apart in a titanic explosion. That cataclysm, observable far across the cosmos, is a Type I supernova. Since all white dwarf stars have approximately the same mass when they explode, experts reasoned that all these supernovae should have about the same absolute magnitude, which should make them good standard candles. They are that, in spite of some brightness differences between them.

Type I supernovae are relatively rare. In the Milky Way Galaxy, there was one in 1006, another in 1572 (which Tycho Brahe saw), and another in 1604 (observed by Johannes Kepler). Not a large sample. To find out whether Type I supernovae make good standard candles and the best ways to use them, astronomers have had to observe them in other galaxies whose distances are known. Fortunately, in terms of their energy output within the visible part of the spectrum, Type I supernovae are the brightest supernovae. Light from the most distant ones observed in the late 1990s has taken more than 7 billion years to reach the Earth.

Type II supernovae, on the other hand, have never been dwarfs. They are exploding giants. The star that explodes in a Type II supernova is definitely well above the Chandrasekhar limit without having had to cannibalize a companion star. When an extremely massive star has used up all its nuclear fuel and can no longer support itself against the pull of its own gravity, it collapses and the resulting explosion is a Type II supernova. Though supernovae of this type are more powerful than those of Type I, they tend not to look as bright to the eye, because so little of their energy comes in the form of visible light. Since they range rather widely in their absolute magnitudes, Type II supernovae are unreliable as standard candles. However, the hope is that measuring the radiation from them, their temperature, and the velocity at which the stars' debris moves apart may provide a way to estimate their individual distances.

What is needed is the discovery of a great many more supernovae in order to put them to optimum use as distance indicators and perhaps find a way to measure their distances independently. Today, there are teams carrying on this search using the Hubble Space Telescope as well as earthbound telescopes in many parts of the world. Their work involves taking photographs of a large section of sky, away from the light of the Milky Way and nearby galaxies, then comparing these pictures with earlier ones of the same region. Computers scan the photographs, subtracting known galactic light, searching for any new light source. If a likely candidate shows up, researchers photograph the same area of sky later to see whether the light has moved. If it has, that probably indicates it came from a cosmic ray or asteroid. Examining genuinely new light sources in detail, astronomers look particularly for spectral patterns identifying those that are Type Ia supernovae—the type most useful as standard candles.

Timothy Ferris in his book *The Whole Shebang* tells of a far more grassroots supernova search. The Reverend Robert Evans of the Uniting Church in Australia compares what he sees nightly through his telescope with his remarkable visual memory of the skies. Evans had discovered twenty-seven supernovae by 1995, a record in the history of astronomy.

Studies of the still expanding debris of old supernovae whose light reached Earth before our time indicate that the 1006 supernova was approximately 5,000 light-years from Earth, and the 1572 supernova about 7,000 light-years away. Tycho Brahe was right in insisting that this one was farther away than the Moon. He called the phenomenon a "nova," but, as modern astronomy makes the distinction, a nova is a less drastic flare-up, usually of a white dwarf star, not an explosion that demolishes an entire star once and for all. Astronomer Fritz Zwicky coined the name supernova in the 1930s.

Among the new techniques that have emerged in the last quarter century, there are several that measure the absolute magnitude of other galaxies, and some of these actually bypass the cosmic distance ladder. One is the Tully-Fisher method, coming from American astronomers R. Brent Tully and U. Richard Fisher. They discovered in 1977 that the absolute magnitude of a spiral galaxy is related to something called its "twenty-one-centimeter line width." Most of the interstellar matter spread throughout a spiral galaxy consists of hydrogen atoms, which emit radio noise at the wavelength of twenty-one centimeters. As a distant galaxy rotates (with some parts of the interstellar matter in it approaching Earth and other parts of it receding from Earth), the Doppler shift causes this spectral line to be blurred. How much it is blurred is directly related to the speed at which the galaxy is rotating. That speed is, in turn, related to the galaxy's brightness. One great advantage of this method is that it is possible to study radio spectra at the twenty-one-centimeter wavelength from extremely faint sources.

Another new technique is the "brightness fluctuation method," which measures the unevenness in the brightness of the surface of the central bulge of a spiral galaxy, or near the center if the galaxy is elliptical. The reasoning is that there should be more apparent unevenness with nearby galaxies than with distant galaxies, because the nearer the galaxy the easier the resolution into stars and the less likely the image will appear as a smooth body of light.

Astrophysicists Rashid Sunyaev and Zel'dovich have discovered a third approach, which measures the distance to faraway clusters of galaxies. They measure the intensity of the cosmic microwave background radiation *through* clusters of galaxies that are emitting

X-ray radiation, and they have found that the background radiation is heated up as it passes through such a cluster. The effect is a "hot spot" in the background radiation. This radiation started out in the early universe with an extremely high temperature, and the more distant the intervening galaxy cluster, the more dense and over-heated the hot spot should be. By studying the hot spot, researchers estimate the distance to the galaxy cluster.

Understanding a fourth method, first suggested by the Norwegian Sjur Refsdal in 1964, requires some background information about gravitational lensing, a phenomenon that was predicted in theory early in the twentieth century but was not observed until much more recently.

Einstein's theory of general relativity links the three dimensions of space and one dimension of time in four-dimensional space-time. According to Einstein, the presence of mass "warps" space-time, and the greater the concentration of mass the greater the warp. To picture this in fewer dimensions, imagine the difference in the ways a bowling ball and a volleyball would create depressions in a trampoline on which they were sitting. The bowling ball is more massive than the volleyball (within a gravitational field like the Earth's, that means it's heavier) and will cause more of a depression. When light travels through space, its path bends as it passes near massive bodies such as planets and stars, or near concentrations of mass such as galaxies and galaxy clusters, in much the same way the path of a tennis ball would bend, veering off to one side, if you attempted to roll it directly past the bowling ball or the volleyball on the trampoline. There is a close-to-home example: The path of light from a star is bent as it passes near the Sun. If an observer on Earth were unaware of the effect, the position where the star appears (detectable during a solar eclipse) would cause that observer to estimate incorrectly the actual position of the star in the sky. (See Figure 7.1.)

It might seem that such distortion would more likely hinder

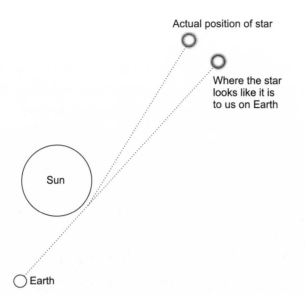

Figure 7.1 A beam of light travels from a distant star, passing near the Sun. The warping of space-time near the Sun causes the path of the light to bend slightly inward towards the Sun. The Sun's brightness doesn't allow us to see such starlight in everyday circumstances, but during an eclipse, if we don't take into account the way the Sun bends the paths of light, it is possible to get a false impression about which direction such a beam of light is coming from and what the distant star's position is in the sky.

than help measurement of distances to faraway objects. Not so. Take, for example, a situation in which light from a remote quasar is traveling across the universe toward Earth. Somewhere between the quasar and Earth is a cluster of galaxies, which is warping space-time around it as such concentrations do. The warp acts like a lens, bending the path of the light from the quasar so that it reaches Earth via not one but two or even several paths. Observers on Earth see two or more images of the quasar—or on occasion a ring of light—instead of a single point of light.

If the cluster responsible for the bending is precisely centered on a direct line between the quasar and the Earth, then the light passing around one side of the cluster will travel the same distance as the light passing around the other. But if the cluster isn't sitting dead center, then the light coming around one side must travel farther than the light coming around the other and it won't reach Earth as soon, since light nearly always travels at the same speed no matter what distance it must cover.

When light traveling from a quasar is lensed by a cluster of galaxies, the angles at which the light reaches Earth allow experts to calculate the extremely tiny fractional difference between the lengths of the paths. However, knowing, for example, that one path is longer than the other by one part in 5 billion doesn't give the actual length of the paths. Quasars have a characteristic that helps at this impasse. They flicker, flaring and fading irregularly over a span of days, weeks, or even years. When the quasar flares, observers on Earth see one image flare before the other, because that particular light has taken the shorter path. From the time delay between the flares, and how much the paths differ in length, it's possible to calculate the distance to the quasar. If, for instance, the time delay is one year and the paths differ in length by one part in 5 billion, the quasar is 5 billion light-years away.

This method doesn't rely on any other steps in the cosmic distance ladder, even though quasars are some of the most distant objects from Earth. There are, in fact, no quasars nearby. They beam their light to Earth like beacons near the limits of the observable universe. So much time has passed since the light we observe left the quasars that we see them as they existed when the universe was only a tiny percentage of its present age. By now, quasars have almost certainly evolved into something quite different from the way they appear to present-day observers. Many astronomers believe they have evolved into the sort of galaxies now situated closer to Earth.

In 1964, when Refsdal first suggested that gravitational lensing might be a clue to the distance of quasars, they were a very recent discovery and still something of an enigma. They looked more like faint stars than galaxies, but they also in some ways resembled gaseous nebulae. They seemed to be very small by cosmic standards, but they were extremely bright, and their brightness varied over days, weeks, even years. All had astoundingly large redshifts.

Sandage was involved in some of the first investigations of these mysterious objects in the early 1960s, along with Thomas Matthews and Maarten Schmidt of the California Institute of Technology. At that time, existing physics was inadequate to make sense of them. Three possible ways of explaining their large redshift all presented problems:

1. The redshift might be caused by a gravitational field. Redshift is the result of the stretching of light waves as the object from which they are coming accelerates away from the observer. Gravity, like acceleration, stretches light waves, which isn't too surprising since even in more familiar situations, gravity and acceleration can feel the same and have the same effect. We experience it when the speeding up and slowing down of an elevator make us feel heavier or lighter, as though the pull of the Earth's gravity has changed. However, in the case at hand, if a gravitational field were causing the redshifts, the objects under scrutiny would have to be so massive and so near that they would disturb the orbits of the planets in the solar system. There was no sign that was occurring, which seemed to rule out possibility number one.

2. They might be stars in the Milky Way Galaxy that have been ejected from somewhere with a force powerful enough to drive them away at a speed necessary for the measured redshift. Examination of their spectra indicated that this was highly unlikely.

3. The best possibility was that the objects are actually very

distant indeed. To cause such a redshift as astronomers were measuring, they would have to be receding at as high a rate as 37 percent of the speed of light. For such a redshift to be caused by the expansion of the universe, they would have to be a vast distance away. How is it, then, that observers on Earth are able to see them as they do? For instance, one of the first studied was 3C273. Its redshift indicated that it is about 2 billion light-years away. Yet it appeared as early as 1895 in photographs taken with optical telescopes of modest size. If it really was 2 billion light-years away, it had to be radiating one hundred times more power than the most luminous galaxies. That also seemed improbable. How large *were* these things?

Their varying brightness was the giveaway. No source of light can flicker faster than radiation can cross it. If it did, the next flicker would begin before the previous one ended and the flicker would appear blurred, not a flicker at all. Radiation can't cross anything at a speed faster than the speed of light. The light output of 3C273 changes substantially in periods as short as one month, which means that most of the light from it must come from a region no larger than the distance light travels in a month.

Using this line of reasoning, researchers calculated that quasars are tiny by cosmic standards, but many have been discovered with greater redshifts than 3C273. For Earthly observers to observe such small objects as they do, at such distances, the objects must be by far the brightest things in the universe, as bright as dozens or even hundreds of galaxies combined. Yet the light from 3C273 comes from a region only one light-*month* across, whereas the light from a galaxy comes from a region of probably some 100,000 light-*years* across or more.

Though they are much better understood now than they were

in the early 1960s, quasars are still mysterious. The source of their power is probably a black hole at the core. Quasars indeed all seem to be incredibly old and far away, some of the most remote objects in both space and time, and with their close relatives, violent BL-Lac objects and blazars, more powerful than any other sources of energy yet discovered.

The Milky Way

Looking up at the Milky Way in the night sky, we are seeing part of the disk of the Galaxy, viewed edge-on (along the plane of the Galaxy). That's the closest Earthly observers can come to "seeing the Galaxy" with the naked eye, and it is a view from inside the Galaxy, not from outside it. Nevertheless, by the mid-1950s, researchers were fairly sure that the Milky Way Galaxy was similar to thousands of other galaxies they *were* able to study at a distance from the outside, and that among them it was midsize. They were able to glean from the study of other galaxies that a galaxy having a mass like the Milky Way has to be one of two types—a gas-free elliptical or a spiral, like Andromeda. The Milky Way is certainly not gas free. That left spiral as the only option.

Like other spirals, the Galaxy must have a pinwheel shape and be composed of a thin disk of gas, dust, and bright, relatively young stars, a central bulge of more densely packed older stars, around which the disk rotates, and a faint halo of even older stars. William Herschel had pictured what we now call "the Galaxy" as a grindstone. Modern astronomers wax even less poetic and compare it to a giant fried egg. The disk is the white of the egg, the central bulge is the yolk. The illustration on page 244 is one of the best recent images of the Galaxy as a whole, showing the disk and central bulge. It came from the COBE satellite. This is not a direct photograph

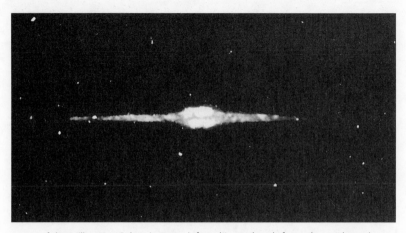

Image of the Milky Way Galaxy in "near-infrared" wavebands from observations of COBE—the Cosmic Background Explorer. *(COBE Science Team, NASA, Goddard Space Flight Center)*

but an image compiled from observations made at infrared wavelengths, a part of the spectrum that is not obscured by dust clouds or more diffuse interstellar dust.

Most descriptions of the Galaxy as viewed face-on, the direction from which it looks like a pinwheel, have the central bulge as circular. It may be necessary to alter that image slightly. Not all spiral galaxies have circular central bulges. Instead, about half appear to have a short bar at the center with the spiral arms beginning at each end of the bar (see comparison in figure 7.2). Observations by a satellite known as the Infrared Astronomical Satellite indicate that stars on one side of the Galactic center are somewhat closer to Earth than stars on the other side. That would be explained if what is out there is a bar of stars set askew of our line of sight (see figure 7.3). Also, the gravitational pull of the rotating bar could be causing the motions of clouds of gas observed near the Galactic center. It may well be that the central bulge is only a short, fat bar, so that viewed from space beyond the galaxy it looks more elliptical than round or rectangular.

Figure 7.2 Comparing the center of a barred spiral galaxy with the center of a galaxy having a circular central bulge.

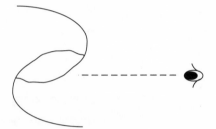

Figure 7.3 The central bulge of a barred spiral galaxy set askew of our line of sight.

Beginning in 1950, William Morgan of the Yerkes Observatory near Chicago pioneered the mapping of the Milky Way's spiral structure. He plotted the positions of two types of brilliant young stars called O and B stars, which are less than about 10 million years old, and found that they are arranged in two lines that run parallel to one another. One of the lines marks the arm in which our solar system lies, now called the Local Arm. The other line is the next arm out, the Perseus Arm. Morgan confirmed these findings by searching for nebulae and studying the stars that light them to estimate their distances. This investigation also revealed a third arm closer to the Galactic center, later dubbed the Sagittarius Arm.

Other optical astronomers, following in the footsteps of Morgan, continued to study O and B stars and nebulae but ran into the problem of the blotchy curtains of dust through which radiation in the optical range can't pass. This barrier is more like a forest in deep fog than an opaque wall, for in some areas it's possible for optical telescopes to penetrate lighter dust—between what in a forest would be bushes and in the Galaxy are black molecular clouds. Nevertheless, optical astronomers hoping to explore near the Galactic center can effectively see little farther than 10,000 light-years, only about halfway there. The view is fifty times better looking in the other direction, from the Southern Hemisphere, away from the center of the Galaxy.

Meanwhile, Dutch astronomer Jan Oort had joined with Australian astronomers in a project to map the spiral arms using radio telescopes in Holland and Australia. Much of the interstellar matter spread throughout a spiral galaxy consists of hydrogen atoms, and these atoms emit radio noise at the wavelength of twenty-one centimers. The study of twenty-one-centimeter emission from hydrogen in the Galaxy is one way of finding out how gas is distributed. Oort reasoned that the motion of the gas would show up as a shift in the wavelength it was emitting and would give him information about the way the Galaxy rotates. Plotting the speed of that rotation against the distance from the Galactic center would enable him to measure the distance to gas clouds and nebulae.

Oort and his colleagues proceeded to map the spiral arms by graphing the intensity of the radiation against the speeds measured from the red- or blueshifts, reasoning that each peak in the graph was a hydrogen cloud and the shift revealed its distance. The resulting map of concentrations of hydrogen gas indicated a pattern of spiral arms, but the map didn't much resemble other spiral galaxies that could be seen more directly. It wasn't until the 1970s that the reason emerged. Hydrogen, it turns out, is not very strongly concentrated in the Galaxy's spiral arms, and some of the peaks on Oort's graph represented gas between the arms instead. Furthermore, the way gas moves and changes its speed upon entering a spiral arm leads to other misleading data. Though hydrogen is not so effective a spiral arm "tracer" as Oort hoped, it did reveal the Outer Arm that lies beyond the Perseus Arm.

In 1976, Yvon and Yvonne Georgelin of France, who also had been following up on William Morgan's work, published a map of the Galaxy based instead on distances to hot nebulae. More recently, their map has been greatly refined by Patrick Thaddeus of Harvard and his colleagues, through the study of molecular clouds, the bushes in the optical forest. As with hydrogen, the motion of such a cloud shows up as a shift in the wavelength of radiation in the radio range of the spectrum—this time from carbon monoxide molecules. Except within a dense molecular ring near the Galactic center—so crowded as to defy mapping of its overall structure—Thaddeus and his team have managed to separate individual molecular clouds along any line of sight by their different velocities, and from those velocities to calculate their distances from the Sun. The result is the most detailed and accurate map yet produced of the spiral arms.

Oort was on the right track to think that the best way to explore the structure of the Galaxy was to study the way things move in it. Most distant clouds of hydrogen or molecules of carbon monoxide have motion toward Earth or away from it, because the Galaxy doesn't rotate in the rigid way a solid structure like a wheel would. What is true in the solar system is also true in a spiral galaxy: The influence of the gravity from the concentration of mass at the center is weaker the farther you are from the center, and that means the outer parts of a spiral galaxy move at slower speeds, just as the outer planets in the solar system do. The result is that distant clouds don't stay continually the same distance from Earth. They have motion as great as 100 kilometers per second toward Earth or away from it, and that motion shows up as a Doppler shift.

In both directions along the Sun's orbit around the Galaxy—
the direction the solar system is headed and the direction from
which it's coming—the nearer gas clouds seem to remain always the
same distance away. Only farther away in those two directions is
there an observable shift. Looking from the solar system in a third
direction, directly away from the Galactic center, there is virtually
no shift. Looking in the direction directly toward the center there is
also no shift nearby, but in the vicinity of the center itself there is
violent noncircular motion of gas. On the near side of the center,
the motion is toward the solar system. On the other side, it's away
from the solar system. Astronomers currently disagree about
whether this movement indicates colossal explosive motion at the
core of the Galaxy or means the central bulge is bar shaped.

Mapping the Universe

The first modern "census" of the universe took place in the 1950s. It
was the National Geographic Society–Palomar Observatory Survey,
which relied on a forty-eight-inch telescope and photographic plates.
Since then, surveying and mapping the universe have become much
more sophisticated.

One question the mapmakers address is: Could Friedmann have
been wrong? Modern cosmology still accepts his assumption that
the universe looks the same in all directions, and that wherever you
are in the universe it will look the same to you in all directions. And
yet that isn't really the way it does look. Certainly not from Earth.
In the sky from the Southern Hemisphere we see the Magellanic
Clouds. We don't see them from the Northern Hemisphere, though
from there we see another distant smudge of light, the Andromeda
galaxy. When we look at the Milky Way, we're looking along the
plane of our Galaxy. We certainly see more stars there than in other

parts of the sky. On this scale, anywhere we go in the universe we will likewise see visible matter distributed unevenly. What is more surprising is that even on the vastly larger scale of a few hundred million light-years, there is still structure, including unimaginably large superclusters, and voids hundreds of millions light-years wide. This is *not* evenly mixed raisin bread dough.

So if you agree with Friedmann that the universe is isotropic (looking the same in all directions) and homogeneous (with matter distributed evenly throughout space), you can't be talking about it on these scales. To find that it is isotropic and homogeneous, you have to look at the universe on a much larger scale yet. How large? In truth, astronomers have not yet found the level at which the universe would all look alike, where it would be impossible to tell one sample of it from another. That has been one of the great surprises of late-twentieth-century astronomy.

The modern picture of the universe on the largest scales emerged in the 1980s and represented a dramatic change from the way it had been visualized before. Margaret Geller, John Huchra, and Valerie de Lapparent, at the Harvard-Smithsonian Center for Astrophysics, decided it was worth following up on preliminary surveys that indicated there might be more structure in the universe than previously suspected. They proceeded to investigate by mapping the redshifts of 1,000 galaxies across one strip of the northern sky. A strip becomes a wedge as you follow it farther and farther into space (see figure 7.4), and this particular pioneering wedge now goes by the name of the Geller-Huchra Wedge.

When they began the project, no one, not even Huchra and Geller, thought it likely to reveal anything particularly startling, and they were in no great hurry to interpret the data. When they did get around to it, jaws dropped. There was an eerie difference in this wedge of the universe from what nearly everyone had been expecting. Here was no homogeneous small-design wallpaper pattern of

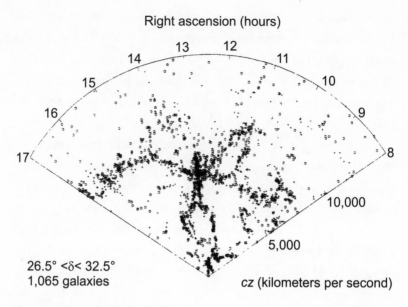

Figure 7.4 The Geller-Huchra Wedge

galaxies with nothing to distinguish one portion of the picture from another. Instead there were huge voids with almost no galaxies in them, bordered by clusters and galaxies strung out like the lights of oceanside towns and cities seen at night from a plane. There was a "Great Wall" of galaxies, a billion light-years long and tens of millions of light-years thick. This was *structure*, to put it mildly! Huchra thought he and his colleagues must have made a mistake. Geller was more willing to relinquish old assumptions. As she quipped in an interview with science writer Timothy Ferris, "I have a strongly held skepticism about any strongly held beliefs, especially my own."

Geller and Huchra proceeded to investigate further by mapping the redshifts in the wedges of space on either side of the original Geller-Huchra Wedge, using better equipment. There had been no error. The awesome structure went on and on.

Other astronomers, including Alexander Szalay, David Koo,

Richard Kron, T. J. Broadhurst, Richard Ellis, and Jeff Munn, have since probed the depths of space by means of "pencil beam" surveys. They limit themselves not to a strip of sky but to an area about half the size of the full Moon. Just as a strip becomes a wedge as it goes deeper into space, a small circle becomes a cone (as in figure 1.4a). A pencil beam survey produces a cone-shaped three-dimensional map that keeps getting larger as it reaches farther distances. The results are plotted in a chart where increasing redshift is the horizontal axis and the number of galaxies discovered at various redshifts is shown as peaks. This method also has revealed the great voids, or superbubbles as some call them.

In December 1995, the Wide Field and Planetary Camera 2 on the Hubble Space Telescope was used to make a different kind of survey. It took 342 exposures over ten consecutive days of a very tiny speck of the sky, 1/140 the apparent size of the full Moon, near the handle of the Big Dipper. This region is relatively uncluttered with nearby stars or galaxies. The camera took separate images through filters for ultraviolet, blue, red, and infrared light. The resulting composite picture is known as the Hubble Deep Field, a narrow, deep "core sample" of the sky, something like the core samples geologists take of the Earth's crust. The Deep Field doesn't reveal the ages of the galaxies it photographed or their distances. Galaxies at many different stages of the universe's history are stacked against one another in the picture.

The Deep Field reached back some 10 billion years and captured an unprecedented view of young, never-before-observed galaxies, some of which are 4 billion times fainter than can be detected by the human eye. Only the Cosmic Background Explorer satellite, when it measured the wrinkles in the cosmic microwave background radiation, looked deeper into space and into the past than this. The Deep Field shows spirals, ellipticals, and a rich assortment of other galaxy shapes and sizes in many stages of evolution. Even though

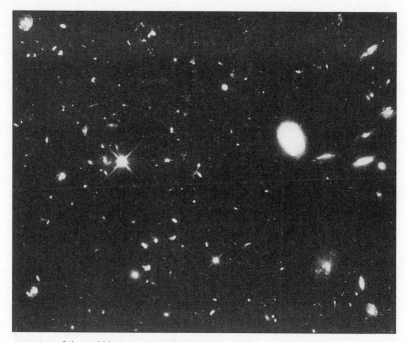

A portion of the Hubble Deep Field. *(NASA/Space Telescope Science Institute)*

the photograph gives no reliable measure of the distances of the galaxies in it, the great number of extremely faint galaxies led astronomers immediately to speculate that some of them formed when the universe was very young. Researchers were soon hard at work measuring redshifts. By the spring of 1997, their calculations had yielded distances for thousands of the faint galaxies as far out as those having a redshift of 1. A redshift of 1 indicates that the light now reaching Earth from that galaxy left the galaxy when the universe was no more than half its present age. Larger redshift numbers mean the light originated even farther back in time. A redshift of 3, for example, means the universe was between 12.5 and 25 percent of its present age.

Studying the galaxies as far out as those having a redshift of 1, researchers interpreted what they found to mean that galaxies with spiral and elliptical shapes probably lead relatively uneventful lives, as galaxy biographies go. It seems that spirals and ellipticals must not change much over billions of years, for the oldest don't look markedly different from those nearby, and the number of them in the distant past was comparable to the number in the universe today. Experts hoping the Deep Field would reveal the formation process of spirals and ellipticals were disappointed. Evidently, to see that, they would have to push the search still farther into the past.

Other types of galaxies in the Deep Field appear to have led more action-packed lives. Their irregular convoluted shapes suggest that galaxy collisions and mergers were far more common in the early universe than they are today, which makes sense, if things were so much more crowded then. Experts have also concluded from the Deep Field that the star formation rate in the universe has declined dramatically during the second half of its history.

For most of the galaxies found at the far limits of the Hubble Deep Field observations, it isn't yet possible to use redshift measurements to calculate distances, because the amount of light from these faint galaxies isn't sufficient for even the largest telescopes to measure their redshifts. Researchers have, however, worked out other techniques. One method is to use other objects as distance calibrators. Fortunately, some galaxies generate powerful emissions in the radio part of the spectrum, detectable at extremely great distances. Present understanding of these radio galaxies has it that this emission comes from their active cores. Also fortunately, many of them are surrounded by other types of galaxies, and measuring the redshift of the radio galaxies allows one to estimate how far away these whole clusters of galaxies are. Some are at redshifts as large as 2.3, and that means the light reaching Earth today left them when the universe was less than 30 percent of its present age.

State-of-the-art astronomy coordinates data from different kinds of telescopes—ground based and space based—observing in different parts of the spectrum. Some distant clusters of galaxies have been studied using the Hubble Space Telescope in conjunction with the most powerful ground-based telescopes, such as the Keck ten-meter telescope in Hawaii, and orbiting X-ray telescopes such as the German ROSAT X-ray Observatory. From this investigation, astronomers have learned that some of the "young" galaxy clusters were probably already extremely massive. Light from them shows that their stars were already mature when that light left them, so the galaxies must have formed much earlier.

Another approach to finding the distances of the faint galaxies in the Hubble Deep Field has been to study the light coming from quasars that are even farther away, looking for evidence in the spectral lines that this light has encountered clouds of gas in halos around galaxies in its journey to Earth. Such detective work has nosed out galaxies at redshifts of 3 and even higher. Another helpful clue has been that all very remote galaxies have a distinctive "signature" in their color. Hydrogen, which is present both in galaxies and in the space between them, absorbs all ultraviolet light shorter than a certain wavelength. The result is that in the spectrum of light from the most distant galaxies there is a cutoff at that wavelength. Using filters, researchers find that a galaxy "disappears" at wavelengths beyond that.

As of 1998, the most up-to-date maps of the universe continued to show clusters of galaxies lining up to form thin walls or filaments enclosing vast voids. Superclusters themselves are parts of supercluster complexes or "walls" or "sheets", enclosing even more enormous empty areas. Someone has commented that if you step back from this picture, you see that the universe is a Swiss cheese. Richard Gott and colleagues at Princeton prefer a different mental image, a sponge. All of the material in a sponge is joined together. So are all the holes. In the universe, the rich clusters of galaxies are more likely to be found at junctures the equivalent of where the pieces of a sponge's material come together. Other theorists think that galaxies, clusters, and superclusters are like a glowing froth on an ocean of dark, invisible matter, similar to the froth that appears among ocean waves. Anyone who has swum in the ocean has seen the voids, filaments, and clusters of foam that form and re-form continually on the swells.

In the existing three-dimensional maps, such as Geller and Huchra's, the largest structures visible are about the size of the volume surveyed, so that in order to get a good statistical sample of these enormous structures and to find out if there are even larger structures, researchers will have to map larger volumes. Two projects are now doing just that. A British and Australian project called the 2DF Galaxy Red Shift Survey uses the 3.9-meter telescope at the Anglo Australian Observatory on Siding Spring Mountain in Australia. By 2000, they hope to have measured a quarter-million galaxy redshifts. Already they are seeing indication of the walls and voids. James Gunn at Princeton has been heading another new project, the Sloan Digital Sky Survey. Even in today's science world, where teams are the norm, SDSS is a remarkable conglomerate. It includes astronomers at the University of Chicago, Princeton, Fermilab near Chicago, Johns Hopkins, the Institute for Advanced Study, the U.S. Naval Observatory, the University of Washington, and the Japan Participation Group.

The Sloan Telescope itself, at the Apache Point Observatory in the foothills of the Sacramento Mountains north of El Paso, Texas, is modest in size but, linked to an impressive array of technology, it is powerful enough to undertake the largest and most comprehensive census of the visible universe that has ever been attempted, a survey of a quarter of the northern sky. Part of the Sloan apparatus consists of CCDs, or charge-coupled devices, silicon chips that convert light from the night sky into digitized images that can be poured onto magnetic tapes and into computers. The project began surveying the sky in the summer of 1998, after nine years of organization, design, and construction, and is expected over the next six years to produce images in five colors of 50 million galaxies, 100,000 quasars, millions of individual stars in the Milky Way, and other assorted items that might turn up. "A field guide to the heavens," Michael Turner (one of the Chicago contingent) calls it. More than a field guide. A three-dimensional scale model of a large part of the universe.

When these projects have succeeded in mapping the universe in three dimensions, the situation may still be like that in the seventeenth century when astronomers had cataloged and charted what was "out there" with greater and greater precision but were waiting for a Newton to come along and discover the dynamics that would reveal *why* things should turn out to be as they are.

Some of the ingredients that will surely go into that understanding are familiar. Gravity and relativity are part of the explanation. Quantum theory—the theory of the very small (atoms, molecules, and elementary particles)—is also certain to be an essential ingredient, because as large as the clusters and superclusters are today, in the early universe the material of which they are made was compressed in an area as small as anything in the quantum world.

Chaos and complexity theorists—who study randomness, the borders between what is random and what is predictable, and the patterns that emerge out of what seems to be chaos—also believe their theories have something to say about the dynamics that formed the large-scale structure of the universe. It would be difficult not to notice the resemblance between the universe on the largest known scales, the "sponge" level, and the graphs and fractals produced in chaos theory. (See figure 7.5.)

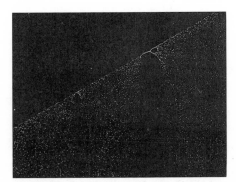

Figure 7.5 Segment of a graph of fluctuations in an animal population, by Robert May—one of the classic early demonstrations in chaos theory.

From Here to Infinity

The best way to put together an overall mental picture of the structure of the Galaxy and the large-scale structure of the universe beyond, as state-of-the-art astronomy describes them, is to take an imaginary tour. At the start, remind yourself that light travels at a rate of approximately 186,000 miles or 300,000 kilometers per second. It takes light 1.3 seconds to travel from the Moon to the Earth, 8.3 minutes to travel from the Sun to the Earth, four hours to travel from the Sun to Pluto, and about 4.3 *years* from the Sun to the nearest star. From four hours to 4.3 years is an enormous leap, but within the range that light from the Sun can reach within seventeen years, there are only about fifty stars.

Ignoring the fact that no human has ever viewed the Milky Way Galaxy from a distance, imagine that your journey begins outside it. You are facing the spiral, with your spacecraft approaching the pinwheel like a moth attempting to fly through the blades of a fan. Your itinerary takes you through the disk, not the central bulge, and you begin by passing through a "layer" several thousand light-years thick, probably consisting of extremely hot gas and faint elderly stars. This region has been difficult to study and, so far, has not been well explored. There is some uncertainty about what you will find there.

Next you penetrate a region called the main disk, about 2,000 light-years in thickness. Here are many of the Galaxy's stars, including the Sun, but your route doesn't take you that close to home. The Sun is in one of the "arms" of the spiral. You travel instead through what looks like a much less densely populated area in between two of the arms. Surprisingly, you don't find as much empty space there as you might have expected. It isn't like flying through a fan and missing the blades, because the arms are nothing so solid and distinct as blades of a fan. The number of stars per cubic light-year is not much less where you are than in a spiral arm. However, if you want more spectacle, you should divert your spacecraft and enter one of the arms, where you will be more likely to encounter extremely brilliant, massive, short-lived stars and glowing nebulae lit by the light of young stars.

Beyond the main disk, you enter the thinnest layer of the galaxy. If the Galaxy were a chocolate-mint wafer, the main disk would be the chocolate and the inner disk you're now entering would be the mint cream layer. This is a disk of gas and dust, only 500 light-years thick, and it is the Galaxy's nursery. It houses the youngest stars and is the birthplace of new stars. If you stayed here awhile and explored it, you'd find that this thin disk of gas doesn't end out where the main disk layers end but instead stretches out beyond that to a distance more than a third again as far from the center of the Galaxy. At its extremes, the gas disk bends like the brim of a hat. At one side of the Galaxy, the edge of the "hat brim" curves upward. On the other side of the Galaxy, it curves downward and then, farther out, back upward.

As you resume your plunge directly through the Galaxy, you find the same layers in reverse until you come out the other side of the Galaxy. For such an enormous Galaxy, the tour seems surprisingly short. That's because compared to its diameter along its plane (looking at it edge-on), the Galaxy is very thin indeed. A chocolate-mint wafer is too thick to be an accurate comparison, not large enough in diameter in relation to its thickness. The Galaxy has more the proportions of a phonograph record.

One thing astronomers now know—to follow up on an earlier attempt to map the Galaxy—is that Harlow Shapley was right to conclude that globular clusters outline the extent of the Galactic halo, though there are also some of them well beyond it. The Omega Centauri Cluster, a globular cluster that Ptolemy cataloged (though not as a cluster), that Halley recognized as a cluster, and about which Herschel spoke in superlatives—"richest . . . largest . . . truly astonishing . . . whose stars are literally innumerable"—does indeed turn out to be extraordinary. It is the brightest, largest, and most massive cluster in the Galaxy, with tens of millions of stars. There are some 100 billion stars in the whole Milky Way Galaxy, many of them congregated in the central bulge, the bright yolk of the fried egg. On a journey directly through the bulge, you might see, or even end up in, a massive black hole that is suspected to lurk at the very center of the Galaxy.

However, saving that adventure for another day, you leave the Galaxy behind, and you begin to notice that among the many billions of galaxies there are in the universe, not all are the same size. They range from about 1 million to 10 trillion times the mass of the Sun, including the extremes. A typical galaxy like the Milky Way is midsize. There is also a much larger structure than galaxies. Moving to larger and larger scales, there are "groups" of galaxies, galaxy clusters, clouds, superclusters, and supercluster complexes or "walls." The terminology suggests a hierarchy, but it is not meant to be understood as anything rigid, and thinking of the universe as a strict hierarchy ignores the rich diversity of its structure.

The first part of your journey beyond the Galaxy takes you

The Andromeda galaxy with two companion galaxies: M32 (the bright dot close to Andromeda on the left) and NGC205 (the bright dot on the right). *(National Optical Astronomy Observatories)*

among the other members of the Local Group of galaxies. This group measures about 5 million light-years across and is rather flattened in shape. "Groups" typically include three to six conspicuous galaxies and a number of smaller, dimmer ones. The Local Group is no exception. The Andromeda galaxy is its dominant spiral with an estimated 400 billion stars. The Milky Way ranks second, and there is a smaller but still impressive spiral called M33. Andromeda has two small companions, M32 and NGC205, both elliptical galaxies. The Milky Way holds court with a retinue including the Magellanic Clouds, which are irregular galaxies, and three dwarf galaxies. The Carina galaxy and the Sextans galaxy are spherical. The Sextans dwarf, discovered in 1990, is a little farther away than the Magellanic Clouds, and the total luminosity of its stars is only about 100,000 times the luminosity of the Sun—less than some single stars in the Milky Way. A more recently discovered satellite galaxy, the Sagittarius galaxy, is the closest neighboring galaxy found so far. First recognized in 1993, it seems doomed to be cannibalized by the Milky Way. It has already lost some of its outer stars to the Galaxy's gravitational pull.

The distance to the Andromeda galaxy is fairly well established at some 2¼ million light-years, but there is still dispute about the distance to M33, the third largest Local Group spiral. When Hubble first measured it in the 1920s, he estimated that M33 was about as far away as the Andromeda galaxy. Sandage reinterpreted Hubble's data on the Cepheids used for that measurement, employing more modern techniques, and concluded that M33 is more like 3 million light-years away, well beyond the Andromeda galaxy. Others who have studied the same Cepheids at infrared wavelengths disagree with Sandage. They estimate that the two galaxies, though farther away than Hubble estimated, are, as he thought, both about the same distance from Earth.

Groups like the Local Group have no particular structure or shape; in fact, they are also known as "irregular clusters," and their galaxies are a hodgepodge of all types. However, that doesn't mean their existence is a random occurrence, with all these galaxies just happening to be passing close to one another on their way to somewhere else. All the galaxies in the Local Group are bound together by mutual gravitational attraction, and all are orbiting a common center of gravity. Andromeda and the Milky Way are, at the moment, approaching each other at a speed of 300 kilometers per second. Slipher's discovery of the blueshift in Andromeda's light was not an error. There could eventually be a head-on collision, but the event is still a few billion years in the future, and the merger of the two galaxies will take another several billion years after that to complete. In the end there will probably be, instead of two spiral galaxies, one huge elliptical galaxy. Then again, Andromeda and the

Milky Way may only circle one another in a polite do-si-do and then move apart again. In 2005, NASA plans to launch the Space Interferometry Mission, a spacecraft carrying an array of telescopes capable of determining, among other things, the exact angle of Andromeda's approach.

You would have to travel several million light-years outside the Local Group to get to the nearest galaxies beyond it. But at this point you can learn more about the large-scale structure of the universe not by traveling farther away from Earth, but by moving to larger and larger scales like "panning out" with a camera.

The Local Group is not far outside the border of the Coma-Sculptor cloud, a large cloud which in turn lies near the outer limits of the Virgo Supercluster. The Virgo Cluster, which is huge compared with the Local Group, is the giant heart of the Virgo Supercluster and about 60 million light-years from Earth. It is an enormous swarm of thousands of galaxies and a lot of hot gas—a "regular cluster" because it doesn't seem to have such a mix of galaxy types as are to be found in motley assortments like the Local Group. Instead, it contains more than 1,000 prominent galaxies centered on a pair of giant elliptical galaxies, with probably many more less prominent galaxies that astronomers haven't yet detected, but very few spirals.

There's little point in even trying to get a sense of the vastness described here. Probably you and I come closest when we are feeling most overwhelmed by our *lack* of ability to conceive of such size and how it compares with familiar distances. The accompanying box gives approximate relative size scales in the universe, but in truth such numbers are beyond human capacity to comprehend. Only for the sake of rough comparison, therefore, here are some figures having to do with the larger scales: A typical "group" of galaxies is a few million light-years across. The Local Group measures about 5 million light-years. A cluster may be 10 million to 20 million light-years in diameter, a cloud some 30 million light-years, a supercluster 100 million or 200 million light-years.

The Relative Scale of the Universe (as calculated in 1995)

The following numbers don't indicate actual sizes, only rough relationships between size ranges, expressed as "powers of 10." To translate: If the exponent is a positive number, it tells you how many 0s are in the number. For example, 10^3 is 1,000 (three 0s), 10^4 is 10,000 (four 0s), 10^8 is 100,000,000 (eight 0s). That means that something designated 10^8 is in the range of ten times as large as something designated 10^7 or a hundred times as large as something designated 10^6. If the exponent is a negative number, it tells how many 0s are in the denominator of a fraction with 1 as the numerator. For example, 10^{-3} is 1/1,000 (three 0s), 10^{-4} is 1/10,000 (four 0s), 10^{-8} is 1/100,000,000 (eight 0s). That means that something designated 10^{-6} is in the range of only a tenth as large as something designated 10^{-5}, a hundredth as large as something designated 10^{-4}, and so forth.

human beings to elephants . . . 1-10	human beings to elephants 1-10
cities . . . . 10^5	limit of visibility of the naked eye . . 10^{-3}-10^{-2}
Moon . . . . 10^6	visible light waves . . . . 10^{-7}-10^{-6}
Earth . . . . 10^7-10^8	molecules and viruses . . . . 10^{-9}-10^{-7}
Sun . . . . 10^9-10^{10}	atoms . . . . 10^{-10}
Earth's orbit . . . . 10^{11}-10^{12}	atomic nuclei . . . . 10^{-14}
solar system . . . . 10^{13}	
galaxies . . . . 10^{20}-20^{21}	
Local Group . . . . 10^{24}	
observable universe . . . . 10^{27}	

Has astronomy discovered the largest structure, or does the universe go on and on to infinitely larger and larger structure? Will the voids turn out to be parts of systems of supervoids? Are the superclusters grouped in clusters of superclusters? *Is* there a level at which the universe is isotropic and homogeneous?

There are tentative answers to those questions. Pencil beam experts probing 6 billion light-years of space (one 3-billion-light-year-long beam pointing out each side of the Galaxy) have found no structure larger than the superbubbles. At extremely great distances, clusters, superclusters, and voids seem to be spread more or less uniformly, with about an equal number in any direction you look. Perhaps it is here, in the ultimate sponge, that there is isotropy and homogeneity.

Another way of approaching the question of at what level the universe is isotropic and homogeneous is to consider the cosmic microwave background radiation. Even though the most interesting news about that radiation has been the discovery of a minuscule *lack* of homogeneity, it *is* remarkably homogeneous, and it is a picture from farther in the past than any other we have. The background radiation is like the smooth-skinned glamour shot showing what a wrinkled dowager looked like in her teens but giving barely a hint of the "structure" in that face today. Ambiguities abound when you start to argue about how far "out" you can peer into the universe and still talk about the universe as it "is," and a dowager universe might well argue that this youthful portrait is "the real me," but it's nevertheless debatable whether this early picture has much to say about whether the modern universe is isotropic and homogeneous.

Astronomy has not been able to reveal an unbroken chronology leading from the era of that radiation through to the present. The universe has a gap in its history like the lost years in the life of someone with amnesia. In terms of the structure of the universe, there is that one window (provided by the cosmic background radiation) into the early universe about 300,000 years after the Big Bang . . . and, next anyone is able to know, there are the voids and supercluster sheets and walls, much closer to the present in time and space. This is the same universe, but no one would guess that from appearances alone. Finding out what happened in the invisible stretch that's missing from the photograph album is one of the challenges facing the next generation of physicists and astronomers.

The Quest
for Omega

{1930-99}

66 When I started, cosmology was very much like philosophy. There was very little chance of measuring something precisely. It's now turning into high precision science. 99
Alexander Szalay

How old is the universe? What is its future? Much of the work going on in state-of-the art physics and astrophysics today focuses on these two fundamental questions. To answer them, it is necessary to know the mass density of the universe—the elusive "omega."

The mass density of the universe means the amount of matter there is per cubic meter, averaged throughout the observable universe. Obviously, this matter is unevenly distributed, at least on normally accessible scales. One way to find out the average density would be to add up all the matter in the universe and then divide by the number of cubic meters in the universe. On the face of it, that would appear to be a ludicrously difficult undertaking.

Never underestimate modern astrophysicists. It is possible to arrive at some estimates. In one method, called "representative sampling," researchers divide the sky into sections of equal size, count the number of galaxies in a section, then multiply the count from that section by the total number of sections. Combined with knowledge about the masses of galaxies, this procedure gives a rough estimate of the total mass of the universe. Studies like the Hubble Deep Field make such sampling increasingly substantive.

Another way to try to find the average mass density is to study the way the universe appears to be working: how fast it's expanding, whether the expansion is speeding up or slowing down, how gravity seems to be affecting different parts of the universe, what forces besides gravity come into play, and how the contents of the universe have evolved over time. This sounds considerably more difficult than counting galaxies and multiplying by sections. It is in fact extremely complicated. Nevertheless, theorists have a formula that they believe shows how the mass density of the universe is related to such questions, an equation that allows them to weigh the answers, one against the other. It is the so-called "equation for omega."

The mass density—omega—affects the future of the universe. If omega turns out to be more than 1 (meaning that there is more than one hydrogen atom per ten cubic meters on average throughout space), the universe will eventually stop expanding and contract. That would be a "closed" universe. If omega is less than 1 (less than one hydrogen atom per ten cubic meters on average throughout space), the universe will expand forever. That would be an "open" universe. If omega is precisely 1, the universe is at the "critical density" that will allow it to expand at precisely the right rate to avoid recollapse, eternally slowing down its expansion but never completely ceasing to expand. That would be a "flat" universe—the universe inflation theory predicts. (See the accompanying box on page 268 for handy reference.)

Why should there be such a tight connection between the mass density of the universe and the fate of the universe? First, mass is the measure of how much matter there is . . . in a planet or star or galaxy or, in this case, in the universe as a whole. Every particle of matter in the universe is attracting every other by means of gravitational attraction. How greatly objects are influenced by one another's gravitational attraction depends on how far apart they are. The closer they are, the more they "feel" one another's pull. So when it comes to the question of whether the universe will eventually contract or whether it will keep expanding, much hangs on how densely or thinly the matter in the universe is spread out. In fact, the mass density very possibly does dictate the fate of the universe.

> Omega is *more* than 1 . . . universe eventually stops expanding and collapses (a closed universe).
>
> Omega is *less* than 1 . . . universe expands forever, thinning out eternally (an open universe).
>
> Omega *equals* 1 . . . critical density; universe expands at precisely the right rate to avoid recollapse (a flat universe, the sort of universe predicted by inflation theory).

Solving the equation for omega requires knowing four numbers, three of which are currently not known with certainty. The four are the speed of light, the cosmological constant (the theoretical constant Einstein suggested and that he hoped would allow the universe *not* to expand or contract), the Hubble constant, and the deceleration parameter. The third and fourth need introduction.

The Hubble constant, or H_0 (pronounced "H naught"), denotes the rate at which the universe is expanding. However, it isn't a direct indication of the speed at which everything out there is rushing away. To give a specific example: If the Hubble constant is 50, that indicates that there is an *increase* in the recession velocity of 50 kilometers per second for every megaparsec of distance from the observer doing the measuring. (A megaparsec is one million parsecs—about 3.26 million light years.) Taking an idealized case and remembering that there are no receding galaxies this close to Earth, if the Hubble constant were 50 and Galaxy A were one megaparsec away from Earth, Galaxy A would be receding at a velocity of 50 kilometers per second. If Galaxy B were two megaparsecs away from Earth, Galaxy B would be receding at a velocity of 100 kilometers per second. A galaxy three megaparsecs away . . . 150 kilometers per second, and so forth. It also means that if these three galaxies were all lined up in a straight line, and if you were in Galaxy A, you would find Galaxy B receding at a rate of 50 kilometers per second in one direction and the Milky Way Galaxy receding at 50 kilometers per second in the other. Notice that this does follow the rule of twice as far, twice as fast, no matter what value the Hubble constant has.

The "deceleration parameter" measures the rate at which the expansion is slowing down due to the gravitational attraction among all the clusters of galaxies. It might seem as if astrophysicists ought already to know what that is. If it's true that when they observe very distant galaxies and galaxy clusters they are seeing them as they were billions of years ago, why isn't it possible to compare the rate at which these are receding with the rate at which nearer galaxies are receding, and find out whether the expansion has slowed down and, if so, how much? Astrophysicists are trying to do just that, but it isn't easy. Some experts have suggested that one reason it's difficult may be that the universe *is* perfectly balanced between the mass density that would allow it to expand forever and the mass density that would cause it to collapse to a Big Crunch. In other words, perhaps the very difficulty of making that determination is a clue that omega equals 1 and the universe is expanding at precisely the right rate to go on forever, always slowing down its expansion but never stopping the expansion and collapsing.

One source of complication is that the mass density of the uni-

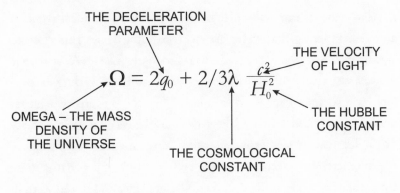

Figure 8.1 The Formula for Omega

verse is changing over time. Unless new matter or energy is appearing on the scene (which the Steady State theory proposed, but most physicists don't believe is happening), things inevitably grow less and less dense in an expanding universe. They thin out.

It is the interrelatedness of these numbers, or values, that's laid out concisely in the equation for omega. It doesn't take much expertise to see that there are relationships, that one thing depends on another. The equation shows precisely in what manner and to what degree they are related. At the risk of sending a great many readers running for cover, here (figure 8.1) is the formula for omega. Consider it a souvenir, something a patient reader is owed for having made it so far with this book. We will *not* proceed to solve it.

Where have researchers got in the process of discovering the unknown numbers in that equation? They know the speed of light. What a pleasure to be able to put in one actual number here! No one yet knows the value for the deceleration parameter or the cosmological constant. There is disagreement over the Hubble constant. Modern scholars find themselves in a situation something like the one their forebears were in when they had Kepler's laws but not

Cassini's and Flamsteed's measurements of the distance to Mars. Here is a formula—but not all the necessary numbers to plug into it.

More Than Meets the Eye

One serious problem in estimating how much matter there is in the universe is that there actually doesn't seem to be enough of it around. In the 1930s, Swiss astronomer Fritz Zwicky discovered that galaxies in the "great cluster" in the constellation Coma Berenices are moving too rapidly, relative to one another, to be bound together by their mutual gravitational attraction. Given the way gravity works, and what can be seen of this cluster of galaxies, the arrangement should be flying apart. Searching for an explanation, Zwicky thought of two possibilities. What appears to be a cluster might instead be a short-term random grouping of galaxies; *or* there might be more to these galaxies than meets the eye or the telescope. In order to provide the amount of gravitational attraction required to bind the cluster together, it would have to contain *much* more. No one was willing to consider the third possibility: that physicists might have made an egregious error in figuring out how gravity operates or in assuming that it operates the same everywhere.

With Zwicky's discovery was born the puzzling notion that it may be impossible to observe more than a tiny fraction of all the matter in the universe. In the years since he first speculated about it, plenty of support—both observational and theoretical—has emerged for the existence of "dark matter." The inescapable conclusion is that for everything to work as it appears to do, there has got to be much more matter in the universe than present technology is able to detect. By some calculations, from 90 to 99 percent of the matter in the universe is not radiating at any wavelength in the entire electromagnetic spectrum. While other pieces of the Big Bang picture fell into place, the missing matter remained a mystery.

There is an example much closer to home than the constellation Coma Berenices: The mass and distribution of observable matter in the Milky Way Galaxy aren't sufficient to account for the way the Galaxy rotates. What would it take to cause the Milky Way to rotate as it does? There should be matter outside the visible disk of the Galaxy, it ought to extend well beyond the edge of the observable disk, and much of it should not be level with the disk but "above" and "below" it. If all that were the case, the rotation would make sense. The suspicion is that the Galaxy must be surrounded by a halo of dark matter that is much larger than the observable mass of the Galaxy. The total diameter of the Galaxy might be four or five times what it's possible to observe in any range of the spectrum. Dark matter might also provide an explanation for the "hat brim" tilt of the Galaxy's thin gas disk.

Dark matter can't be investigated directly, only by watching how it affects other things—more specifically, what its gravitational effect is on other matter and radiation. Sometimes it gives its presence away by the manner in which it bends the paths of light. Such paths through space-time are bent by the presence of massive objects ("benders") such as stars, planets, galaxies, and galaxy clusters. This happens regardless of whether the benders are themselves detectable at any wavelength. When the distortion is too great to be caused by the observable matter in the bender, or when there is no observable bender at all, researchers know they are not observing everything that's out there between them and the background. They suspect the presence of dark matter.

The mystery of dark matter lies at the heart of the problem of measuring the age and the future of the universe. Calculating roughly whether there is sufficient observable matter in the universe to produce the gravitational attraction necessary to keep the universe at critical density, omega-equals-1, shows that the amount of matter observed directly with present technology falls far short. But the discussion doesn't end there, because dark matter does exist and because no one is yet certain how much there is or what it is.

Big Bang theory in its most straightforward form has it that even a microscopic deviation from omega-equals-1 would have caused the universe very early on to recollapse or would have made it expand so rapidly that stars could never have formed. Inflation theory has proposed a solution to that fine-tuning problem, but the question right now is: *Is* the universe really all that fine-tuned? It seems that things, including people, couldn't exist as they do if it weren't, but it isn't actually obvious how it *is*. No measurement of existing mass density comes anywhere near critical density, which means that the universe should have expanded too fast for stars to form. It didn't. What is it we don't know yet?

Candidates for dark matter range from still hypothetical mysterious exotic particles to black holes a billion times more massive than the Sun. Planets, dwarf stars too dim to have been observed, massive cold gas clouds, comets and asteroids, and an assortment of dead or failed stars make up a broad middle ground of possibilities. Some physicists insist on tossing in a few copies of the *Astrophysical Journal*.

In 1998, hard-to-detect particles known as neutrinos moved to the short list. The existence of neutrinos was first suggested in 1930 by Wolfgang Pauli as a way to explain a mysterious loss of energy in some nuclear reactions, but it was not until 1956 that observations by Frederick Reines and Clyde Cowan at the Los Alamos National Laboratory in New Mexico confirmed their existence outside of theory. Neutrinos remain notoriously difficult to study. They rarely interact with any kind of matter. A neutrino can pass through a piece of lead a light-year thick without hindrance. Clues to their existence come on those rare occasions when a neutrino does happen to collide with an atom, but even then the evidence is indirect.

Whether neutrinos have any mass at all has been in question, and of course if they have no mass they cannot be contributing to the mass density of the universe. There have been a number of claims in the last few years of the discovery of neutrino mass, but much stronger evidence came in June of 1998, from a team of Japanese and American physicists at an observatory in Takayama, Japan.

Their detector was a tank the size of a medieval cathedral containing 12.5 million gallons of ultrapure water, inside a deep zinc mine one mile inside a mountain. One of the ways neutrinos are produced is when cosmic ray particles from deep space slam into the Earth's upper atmosphere. Experimenters hoped to compare neutrinos that came from the upper atmosphere directly over the detector (a short distance) with those that were coming up from under the detector after having passed through the Earth (a long distance). Neutrinos from both sources, moving through the water, would occasionally collide with an atom. The result of such a collision is a scattering of debris, and the particles of that debris race through the water creating cone-shaped flashes of blue light called Cherenkov radiation. The light is recorded by 11,200 twenty-inch light amplifiers that line the inside of the tank. Researchers analyze the cones of light, finding the proportions of different sorts of neutrinos coming from each direction and attempting to determine whether the neutrinos, which come in three types, change type on their journey from the upper atmosphere. If neutrinos can make this change, that means they must have mass.

Yoji Tkotsuka, leader of the team and director of the Kamioka Neutrino Observatory, the site of the detector, announced that the evidence was strong for neutrino mass. "We have investigated all other possible causes of the effects we have measured," he reported, "and only neutrino mass remains."

Calculations based on these findings show neutrinos might (not everyone agrees they do) make up a significant part of the mass of the universe. Not that a single neutrino amounts to much. The mass of a neutrino turns out to be almost infinitesimal—about $1/_{500,000}$ of the mass of an electron. Nevertheless, neutrinos pack con-

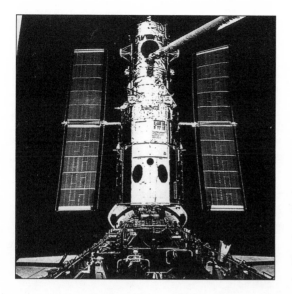

The Hubble Space Telescope, linked to the Space Shuttle *Endeavour* during the December 1993 repair mission. *(NASA/Space Telescope Science Institute)*

siderable clout by dint of their numbers, for there are about three hundred of them in every teaspoonful of space. They outnumber other particles in the universe by a billion to one. In fact, the discovery of that tiny mass by the team in Takayama adds considerably to the mass density of the universe, by some calculations more than doubling it at one stroke.

As promising as this might appear, the discovery that neutrinos have mass can't account for all the missing matter that calculations show ought to exist. And though the combined mass of all those neutrinos may be enough to slow the expansion of the universe, it isn't likely to be enough to stop it or turn it around. The discovery of neutrino mass hasn't revealed the future of the universe.

A Glitch in Time

In the mid-1980s, a panel of astronomers reviewing plans for use of the Hubble Space Telescope decided that the determination of an absolute distance scale outside the Galaxy and the discovery of the expansion rate of the universe—the Hubble constant—should be among the highest priority projects undertaken by the telescope. A team of astronomers from the United States, Canada, Great Britain, and Australia, led by Wendy Freedman of the Carnegie Observatories in Pasadena, California, received the largest allocation of time on the Hubble Space Telescope for a period of five years. The Extragalactic Distance Scale Key Project, as they call their work, involves trying to determine distances to nearby galaxies more accurately than ever before. These galaxy distances will then form the underlying basis for a number of other methods that can be applied at more remote distances, making possible several independent measurements of the Hubble constant.

Wendy Freedman is a native of Toronto, Canada, and one of her most vivid childhood memories is of a trip with her father to northern Canada, where they watched the stars and he explained to her how long it takes their light to reach us on Earth. When Freedman entered the University of Toronto in 1975, she intended to study biophysics, but soon she switched to astronomy. She got her doctorate in astronomy and astrophysics from Toronto in 1984, then received a Carnegie Fellowship at the Carnegie Observatories, and in 1987 was the first woman to join the permanent faculty there.

At the heart of Freedman's Extragalactic Distance Scale Key Project lies the effort to measure Cepheid distances to twenty galaxies with the Hubble Telescope. These distances are then expected to provide an absolute scale for other methods that give only relative distances (Type Ia supernovae, Type II supernovae, the Tully-Fisher relation, and surface-brightness fluctuations).

In 1994, Freedman and her team were attempting to measure more precisely the distance to the center of the Virgo Supercluster. They found twenty Cepheids in the spiral galaxy M100 in the Virgo Cluster, at the core of the supercluster, the first sure identification

Wendy Freedman. *(Courtesy of Wendy Freedman)*

of Cepheids that far away. The Hubble data indicated that these Cepheids are approximately 56 million light-years from Earth. Interestingly, that was nearer than earlier estimates put the center of the Virgo Supercluster. From this new distance measurement and M100's recession velocity (learned from its redshift), Freedman and her colleagues calculated a new value for the Hubble constant, about 80 kilometers per second per megaparsec. Experts, led by Allan Sandage, had previously calculated that its value was about 50. Thus began one of the most heated debates in modern astronomy—either an extremely significant discussion or much media hype about nothing, depending on which side of the issue you stand.

Freedman's announcement came as a shock. The Hubble findings were an embarrassment. Deciding between values of 50 and 80 was not mere nit-picking: If the universe is expanding so much more rapidly than previously thought, it follows that less time has elapsed since the Big Bang than the 10 billion to 20 billion years most experts had settled on. Depending on the density of matter in the universe, a Hubble constant of 80 means the universe must be only 8 to 12 billion years old, probably nearer to 8 billion.

It isn't uncommon in science for new findings to challenge earlier thinking, sometimes eventually undermining what nearly everyone has been assuming was virtually unassailable scientific knowledge. But this challenge—if it wasn't just a glitch in observation or interpretation blown out of proportion—was one of the most disquieting so far in the twentieth century, for astronomers were fairly certain, based on what they considered sound understanding of nuclear physics and the rate at which hydrogen converts to helium in stars, that some of the oldest stars in the Milky Way are 14 billion years old, probably even older. The universe can't be younger than the stars in it.

The discrepancy made the front pages of newspapers all over the world. Scientists ground their teeth. It's difficult not to notice that the attitude of the science community in the twentieth century is sometimes startlingly reminiscent of the attitude of the Church in Galileo's day. There is a reluctance to let the public know about anything that might undermine simple faith—in science, this time around. Since the future of astrophysics and astronomy depends on massive public spending, these branches of science do have an enormous stake in maintaining their credibility. Researchers are concerned that if public opinion is to favor continuing support for this wondrous but expensive adventure now that the old Cold War rivalry is history, there should be no announcements indicating that tax money is buying nonsense!

Freedman and her team are young astronomers. Those whose numbers they were questioning are some of the most highly—and deservedly—respected older members of the astronomy community. Sandage had spent the good part of a lifetime developing new measuring techniques and making careful observations to arrive at the Hubble constant value of 50. But iconoclastic as the team's announcement was, it didn't come entirely out of the blue, nor was such a conundrum unprecedented. As far back as 1929, Hubble himself calculated the Hubble constant to be 500 kilometers per second per megaparsec, making the universe younger than geologists knew the Earth was. Baade refigured H_0 at 250. Sandage and Tammann reduced it still farther to 180, then to 75, then (in the mid-1970s) to 55 plus or minus 10 percent. These corrections succeeded in making the universe old enough to allow for the formation of even the most ancient stars and globular clusters, but not before discussions had taken place that resembled those that now followed Freedman's announcement. Nor had that lowest value for the Hubble constant previously gone unchallenged.

In the late 1970s and 1980s, when Sandage had settled on a Hubble constant close to 50 and an age of the universe of 15 to 20 billion years, Gerard de Vaucouleurs of the University of Texas took serious issue with those numbers. Shortly before Freedman made her findings public in October 1994, there were other studies whose results implied that current estimates of the universe's expansion rate and age might be headed for another revision. A team led by Robert Kirshner of the Harvard-Smithsonian Center for Astrophysics, using the Cerro Tololo Inter-American Observatory in Chile, measured the expanding debris from five supernovae and judged the universe to be from 9 billion to 14 billion years old. But Freedman's team's calculations, based on data from the Hubble Telescope, were more compelling than any of these other challenges to the older numbers.

Astronomers leave a wide margin for error in calculations such as these, and the immediate temptation is to wonder whether the numbers are sufficiently fuzzy to allow the universe to be just barely old enough and the oldest stars just barely young enough. However, stars didn't pop into existence the instant the universe began. Estimating that stars are the *same* age as the universe would be unsatisfactory. There must be a cushion of at least a billion years after the beginning to leave sufficient time for them to form. The leeway in Freedman's numbers and in current estimates of the age of the oldest stars is not enough. For reference: A Hubble constant of around 50 indicates an age for the universe of around 15 billion years; a Hubble constant of around 70 or 80, a much younger universe—about 10 billion years or less.

A negative reaction to the Hubble team's announcement came almost immediately, and not unexpectedly, from Sandage, whose office was right down the hall from Freedman's at the Carnegie Observatories. Sandage had served under Hubble himself here when the observatory was known as Mount Wilson. According to Sandage, the glitch was being grossly overpublicized and its importance exaggerated by the media. This was no more than the sort of routine apparent conundrum scientists grapple with all the time. There were plenty of possibilities of error in the Hubble team's results: in their measurements of the apparent magnitudes of the Cepheids, for instance, and in their assumption that the galaxy where these Cepheids are is actually in the center of the Virgo Supercluster. Perhaps instead it is in the foreground, nearer to the Milky Way. Arguing for that is the fact that M100 is a spiral galaxy, and it is elliptical galaxies, not spirals, that are more commonly found in the centers of clusters like Virgo. Freedman countered that her team had already taken that possibility into account in assigning a wide margin of error to their estimate. What's more, as they had also reported in their original paper, the relationship of Virgo to another distant cluster, the Coma Cluster, had made it possible to "step out" to there and calculate the Hubble constant at that distance—a calculation that gave the same result.

The question also arose whether the rate at which Virgo is

moving away is a dependable indicator of the recession rate of the universe as a whole. Tammann reported that his studies indicated that Virgo is actually moving away more rapidly than the rest of the universe. Here, again, was the perennial problem of sorting out the actual "Hubble flow" from all the other movement that's going on among and within clusters and superclusters. How to extract from this complicated picture the part of all that motion that is directly attributable to the expansion of the universe? Any sample of the universe is likely to give a faulty reading unless it is an extremely large sample indeed. No one knows for certain how large a sample that would have to be.

Freedman and her team hadn't claimed that their result settled once and for all the value of the Hubble constant, but neither were they convinced by the opposition. Sandage's own measurements had recently been challenged on another front. He had been using Type Ia supernovae for making his distance measurements. Some of the measurements that gave Sandage and his collaborators a Hubble constant of around 50 relied on the assumption that these all reach the same maximum brightness and are good standard candles. However, Mark M. Phillips, an astronomer at the Cerro Tololo Inter-American Observatory in Chile, had recently found that not all Type Ia supernovae do have the same brightness characteristics. Brighter ones appeared to occur in spiral galaxies or galaxies with many bright stars. Phillips had developed a technique for analyzing the light curves (how the supernova brightens and dims) to recognize these differences and make allowances for them, but, as of 1994, Sandage had not corrected his data. Robert Kirshner and his colleagues at the Harvard-Smithsonian Center for Astrophysics *had* corrected theirs, and their estimate for the Hubble constant had risen from 55 to around 67.

At the American Astronomical Society meeting in January 1995, just two months after Freedman's announcement, the new measurements and the controversy they'd stirred up about the value of the Hubble constant and the age of the universe took center stage. Phillips and Mario Humay, Phillips's colleague in Chile, were there reporting their measurement of twenty-five supernovae, some as far away as 1 billion light-years. Compensating for the differences in maximum brightness that Phillips had discovered, they'd arrived at a Hubble constant of 60 to 70, in the middle range between Sandage's measurements and those of Freedman's team. Freedman announced that the Hubble Telescope had now measured distances to forty more Cepheids in M100 and distances to two other galaxies in the Virgo Cluster, M101 and NGC925. The new data were consistent with her team's earlier results.

Eight months passed, and in September 1995, Nial Tanvir at Cambridge University, with colleagues at Durham University in England and the Space Telescope Science Institute in Baltimore, estimated—based on fresh Hubble observations of Cepheids—a distance of 38 million light-years to the M96 galaxy, in the direction of the constellation Leo. From this they inferred a distance to the much more remote Coma Cluster. The team's calculation gave the universe an age of 9.5 billion years, give or take a billion.

In March of 1996, a year and a half after Freedman's initial announcement, Sandage and colleagues had rallied and were ready to report the results of their ongoing supernova study. In 1990, there had been a Type Ia supernova in the galaxy NGC4639. Sandage's team had been observing the light curve of this supernova since 1992. What was particularly significant about this investigation was that the Hubble Telescope had been able to see individual stars in the same galaxy and twenty of them were Cepheids. From their brightness, Sandage's group had been able to calculate the distance of the NGC4639 galaxy as 82 million light-years, and from that they knew the absolute magnitude of the Type Ia supernova in a way that didn't depend on comparing that supernova's brightness with the brightness of other Type Ia's. Applying this fresh knowledge to previous measurements of the apparent peak brightnesses of six other Type Ia supernovae, Sandage recalculated their distances and came up with a Hubble constant of 57, in the range he had been insisting on all along. Sandage told an interviewer, "We believe that this marks the end of the 'Hubble wars.' "

Freedman and her colleagues were not won over. Sandage's new results, compelling as they seemed, didn't make the team's own Hubble findings go away or point up any flaw in their calculations. The Freedman team's numbers did come down a little. At the Princeton "250th Birthday Conference" in June 1996, she reported a value of 73, based on a combination of the distance measurements to Cepheids, the study of Type Ia and Type II supernovae, the Tully-Fisher relation, and surface-brightness fluctuations. Freedman said that with so much data accumulating so rapidly, the debate might well be settled in the next three years. Sandage says he is sure that it will finally be settled close to his own number, but perhaps not until well into the next century, too late for him personally to savor his victory.

What about the age of stars? In the winter of 1996, study of some of the most distant galaxies ever observed showed them to be as much as 14 billion years old, shoring up faith in earlier calculations. But in the late summer and fall of 1997, physicists at Case Western University in Ohio, led by Lawrence M. Krauss, reexamined the age of some of the oldest and most distant stars, using measurements from the *Hipparcos* satellite. They recalculated the age of globular clusters previously thought to be as much as 15 billion years old or even older. *Hipparcos*'s measurements revealed that these globular clusters are farther away than earlier estimates had put them, and so, in order for them to *appear* as bright as they do, they must also be brighter than previously thought. If they are brighter, that means they are burning faster and have evolved more quickly,

284 MEASURING THE UNIVERSE

making them younger—perhaps 11 billion rather than 15 billion years old.

Hipparcos cannot directly measure the distance to these globular clusters. They are in the Milky Way's halo, outside the main disk, too far away for parallax measurements, even with *Hipparcos*. Instead, researchers used *Hipparcos* to measure the distance and brightness of other stars and compared them with stars of similar composition in globular clusters. Catherine Turon of the Paris-Meudon Observatory, who along with others calculated 12.8 to 15.2 billion years for the age of a globular cluster known as M92, admits there are difficulties with such measurement: The stars used for comparison are often dim stars that contain almost no metals or other heavy elements. Getting models adapted to such objects is problematic. Processes such as fast rotation or metals sunk out of view into the star could skew the conclusions. Michael Perryman, a project scientist for *Hipparcos*, is even more skeptical. *Hipparcos*'s own data have shown that some of the stellar models are spectacularly wrong. Needless to say, these new calculations of the age of stars did not put the age-of-the-universe paradox to rest.

However, also in the autumn of 1997, astronomers using the Hubble Telescope found stronger indication that old assumptions about the history and ages of globular clusters were not unshakable. They watched the collision of two galaxies called the Antennae and observed at least 1,000 clusters of newborn stars forming from giant hydrogen clouds in the center of the merging galaxies, showing that not all globular clusters are necessarily among the oldest objects in the universe. At least some are emerging out of more recent galactic collisions. But Freedman points out that though not all globular clusters are as ancient as previously thought, a great many in our Galaxy *are*. Furthermore, continuing studies of faint galaxies in the Hubble Deep Field argue for an older universe, for they show that some elliptical galaxies were already well advanced in years at a redshift of 1.2. It doesn't help that *some* things are younger than previously thought.

Sandage, Freedman, and their associates are exploring the frontiers of their field. This is very recent and ongoing science. The Cepheids in the Virgo Cluster that stirred up the controversy couldn't be seen at all before the Hubble Space Telescope's faulty optics were corrected in December of 1993, less than a year before Freedman's team's discovery. Some of the measuring techniques being used are barely past the experimental stage, yielding data whose implications no one fully understands. It would be the stuff of tabloids to declare anything settled. Those who are uncomfortable when science yields paradoxes rather than certainties will have to go on being uncomfortable for a while.

Einstein's "Blunder" Revisited

Suppose the Hubble constant does end up indicating that the universe is younger than some of its stars. It may seem relatively easy to think of the rapidity with which the universe is expanding as being the result only of a two-way tug-of-war between gravity (which is working to make the universe contract) and the expansion energy resulting from the Big Bang (which is working to make it expand). There may, or may not, be another player involved, and that player is Einstein's old "mistake."

Popular science books and articles habitually describe the cosmological constant as a repulsive force that might counter the effect of gravity, but the cosmological constant can actually work either way. If it is a positive number, it will indeed counter gravity, joining in the struggle on the side of expansion. However, if it is a negative number, the effect will be to weigh in on the side of gravity. If it is zero, it will do neither. To put that another way, imagine yourself facing a dial. Make the arrow point to zero; then the cosmological constant will have no effect at all. Move it to the minus side of zero, and the farther you turn it, the more it will contribute to the contraction of the universe. If you move it to the plus side of the zero, the farther you turn it the more it will contribute to the expansion of the universe.

Even this slightly more sophisticated view of the cosmological constant does not begin to do justice to the complications involved when physicists and astrophysicists play with its value. The cosmological constant can seem to be working both ways at the same time, allowing one to have one's cake and eat it too. However, what seems a contradiction is not, because of the way the cosmological constant fits into the equation for omega.

Theories of quantum mechanics—the study of the very small (atoms, molecules, and particles)—have it that everywhere in the universe particles are spontaneously popping into and out of existence. Their life spans are unimaginably short. Nevertheless, "empty space" seethes with this energy, and "empty space" does not mean only what is out there dark and remote between the stars. This quantum energy fills the enormous amount of empty space within the atoms that make up chairs, tables, human bodies, and all other things familiar and unfamiliar. "Emptiness" is full of energy. Theory suggests that the energy of the cosmological constant might be this energy of virtual particles that wink in and out of existence at all times and everywhere in the universe.

For Einstein, the cosmological constant was only a mathematical device, and not long after he put it into his equations in order to avoid the implication that the universe must be either expanding or contracting, he decided it had been a mistake—for, of course, the universe *is* expanding. After visiting Hubble at Mount Wilson, in 1931, Einstein rejected the whole idea of a cosmological constant, calling it "theoretically unsatisfactory anyway." But it didn't go away.

Lemaître in particular enjoyed fooling around with it and adjusting its value, discovering that by fiddling with this theoretical dial he could construct universes that started out very slowly and then sped up, or universes that started out fast and then slowed down, or universes that began expanding, stopped, and then expanded again. Something like that stop-and-start version was evoked in the 1940s as a possible remedy when new discoveries indicated that the universe was younger than the solar system. When it then turned out that the Hubble constant had been overestimated, the cosmological constant wasn't required after all and was packed away once again.

In 1948, researchers detected the effects, on atoms, of the vacuum energy decreed by quantum mechanics. But no one went on to study its possible influence on the universe as a whole until nineteen years later when Zel'dovich, the Soviet theorist, realized that this vacuum energy would enter into Einstein's equations in just the same way that the old cosmological constant did. Before long it became evident that if Einstein had been right about mass/energy causing space-time to curve, and if all this vacuum energy really does exist, then the vacuum energy ought long ago to have curled up the universe into a tiny ball or something even smaller, or else driven the expansion so that even atoms—much less galaxies—could never have formed. Even by making the cosmological constant extremely small, Zel'dovich couldn't show how the universe could have turned out to be the way it is. So it seemed the value must be zero, as most theorists since have been assuming. That zero does not mean there is no vacuum energy, only that by some truly remarkable coincidence, all the positives and negatives in that vacuum energy cancel out exactly.

In current debates, the cosmological constant is still with us, hovering like a ghost in the equation for omega. If its value is zero, scientists could call it a wash and forget it, but the symbol for it would still be sitting there. Even before the recent discoveries that the universe may be expanding faster than previously supposed, late-twentieth-century astrophysicists had been feeling once again an itch to reach for the cosmological constant dial. There was a possibility it might offer solutions to some intractable problems such as the still missing dark matter.

Nearly everyone was approaching the idea very cautiously, having been burned before. John Noble Wilford commented in a *New York Times* article that one thing that makes physicists particularly uneasy about assigning the cosmological constant a value other than zero is that this reminds them too much of the way medieval astronomers designed increasingly complicated celestial mechanisms to explain the planets' motions, in order to preserve their beloved Earth-centered Ptolemaic universe. As was true with those mechanisms, there was nothing to indicate that a cosmological constant value other than zero is wrong. But there was also nothing to indicate it is right. The strongest argument for it was that it allowed physicists to cling to theories in which they had a vested interest! Being able to turn the cosmological constant dial to a number (negative or positive) of their choice that ends up supporting the currently favored version of the Big Bang theory was just a little too easy and allowed for too much leeway. With a friend like the cosmological constant, did a theory need enemies?

In 1990, Michael Turner of the University of Chicago and Fermi National Laboratory proposed a recipe to add up to critical density: 5 percent ordinary matter, 25 percent cold dark matter (including both invisible and "exotic" types), 70 percent the cosmological constant or something like it. According to Turner, the energy of the cosmological constant could compensate for some of the missing mass and serve as an additional brake on cosmic expansion, balancing things out in such a way that the universe would eventually neither collapse nor expand into an ever darker, thinner, colder infinity, but instead perch for all eternity on that highly desirable knife edge between the two. The lower density of matter that such a cosmological constant value would allow might be an added boon to theorists, making it easier to explain how matter congealed into such enormous structures as the Great Wall of galaxy clusters.

After Freedman's team's discoveries in late 1994, physicists began to consider much more seriously these suggestions that the cosmological constant might not be zero. Adjust the dial, and the cosmological constant's energy, over time, could change the rate at which the universe expanded. If expansion was slower when the universe was young, that would have given more time for stars and large structures to develop. Later, the energy of the cosmological constant could have influenced the expansion to speed up. The current measurements of the rate of expansion, by Freedman and others, might be only measurements of the *present* rate of expansion, and unreliable indicators of the age of the universe.

However, though the notion of a cosmological constant that wasn't zero became more and more tempting in terms of explanatory power (in other words, useful to explain what was going on), that was not the same as having observational evidence that the value was something other than zero. The first hint that there might be such evidence came in 1996, though not from direct observation.

The age-old method of testing alternative ideas and making up for insufficient evidence by using mathematical simulations came into its heyday with the advent of supercomputers. In the 1980s and 1990s, as never before, it was possible to feed in some observed and assumed conditions in the early universe and find out what this might lead to after billions of years. In 1996, an international team headed by Carlos S. Frenk of Durham University in England, using Cray supercomputers at Munich and Edinburgh, ran simulations to find out whether temperature fluctuations observed in the early universe could have led from a Big Bang fireball, where everything was almost uniform, to today's universe of galaxies, clusters and voids.

The starting point for the simulations was the universe as it is thought to have existed 300,000 years after the Big Bang. That was when the cosmic microwave background radiation originated (the radiation that Penzias and Wilson detected in 1964). In 1992, George Smoot and his colleagues had discerned tiny energy fluctuations—"wrinkles"—in this radiation. Frenk's team simulated the growth of those initial wrinkles. The results supported the possibility that the cosmological constant should indeed be summoned from limbo.

Frenk and Simon White at Munich (assisted by Adrian Jenkins, Frazer Pearce and Joerg Colberg) ran four different simulations. One of the ways they differed from one another was in their estimates of the mass density of the universe. A second difference was in whether or not Frenk and his colleagues allowed the cosmological constant to be other than zero. Of the four (see figure 8.2), the one capable of producing the universe as we know it today was the model in which the mass density was only 30 percent of what experts think would be needed to produce omega-equals-one, or critical density, *and* in which Frenk also factored in a cosmological constant other than zero.

In this simulation, the expansion rate changed over time and was slower in the early universe than it is today. The computers demonstrated how the wrinkles might have attracted surrounding matter. Lumps of matter, according to the simulation, collapsed onto themselves and grew larger by merging with other lumps, eventually forming a complex filamentary network of large, twisting ridges surrounding vast empty regions. Gas and dark matter flowed along these filaments. Where the filaments intersected, galaxies and

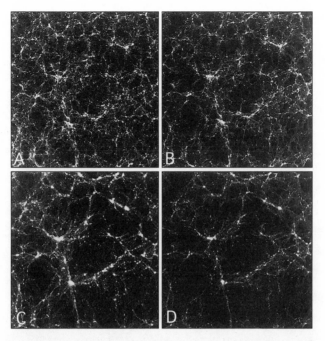

Figure 8.2 Computer simulations from Carlos S. Frenk and his colleagues. C. most re-sembles the universe today. It is based on a mass density only 30 percent of what we think would be needed to produce omega-equals-one, or critical density, and Frenk also factored in the cosmological constant. A. and B. are much less successful models based on greater density. D. is based on 30 percent density, without the cosmological constant.

galaxy clusters formed. In the simulation, the last few billion years don't show much gross alteration, for the universe is expanding fast and the mass density is too low for the large structures to change very much.

No simulation, by itself, can provide the answers to the questions about expansion rate and age, the cosmological constant, and the missing mass. But Frenk, in an interview with the *New York Times*, argued that his simulations do point out strengths and weaknesses in several theoretical models and "give us greater confidence in what are, you might say, the best-buy models of the universe."

Frenk's results were in line with those of more modest simulations by Jeremiah Ostriker of Princeton and Paul Steinhardt of the University of Pennsylvania, and with models developed by James Peebles of Princeton.

The possibility raised by the simulations that omega does not equal 1 echoed some recent observational evidence: Studies of the spectra of galaxies in the X-ray range had been raising questions about proportions of ordinary matter and exotic dark matter. Also, the discovery of ever-larger galactic superclusters seemed unexplainable if omega does equal 1. On the other hand, simulations by Joel Primack at the University of California, Santa Cruz, and scientists at New Mexico State University seemed to rule out the cosmological constant. "No one," said Peebles, "should start collecting bets."

That was how things stood when, at the January 1998 meeting of the American Astronomical Society, Saul Perlmutter of the Lawrence Berkeley National Laboratory in California made an announcement that many of his colleagues predict will be as significant a watershed as Slipher's report that nearly all galaxies are receding from Earth. A team called the Supernova Cosmology Project, which Perlmutter heads, had been studying supernovae to find out whether the expansion of the universe is slowing down. Perlmutter reported that not only does the expansion show no signs of slowing down, but there actually seems to be observational evidence that it is speeding up.

Since childhood, Perlmutter had always been deeply interested in the most fundamental questions of how the universe works. As an undergraduate at Harvard and working toward a Ph.D. at Berkeley, he'd become increasingly convinced that serving on teams involving hundreds of participants—as is common in modern world-class particle physics—would give a young physicist little chance to shape the research. How else to ask the fundamental questions?

Saul Perlmutter. *(Ernest O.
Lawrence, Berkeley National
Laboratory)*

Perlmutter decided to try astrophysics, and that is where he is
today, shaping research that may indeed lead to the answers to his
questions. But his experience in fundamental physics has inclined
him to be more patient and less resistant than some of the astronomy culture can be to projects that take years of single-minded
pursuit to complete.

Ten years ago, Perlmutter began what promised to be a lengthy
and difficult endeavor indeed—and at the outset something of a
gamble—using distant supernovae as mile markers to measure
trends in cosmic expansion. When preliminary results satisfied
him and others that supernovae could be used effectively for such
measurement and that available technology should be up to
the task, Perlmutter and his team dug in for a long-range investigation.

The most distant supernovae that Perlmutter's team had discovered by January 1998 were some 7 billion light-years away, meaning that by the time their light reached telescopes on Earth, 7 billion

years had passed since the stars exploded in supernovae. By now that light is feeble, redshifted by the expansion of the universe. The Supernova Cosmology Project involves comparing the feeble light of these distant supernovae with the light of bright nearby supernovae to determine how far the faint supernovae light has traveled. The distances combined with redshifts of the supernovae give the rate of expansion of the universe over its history, allowing researchers to determine how much the expansion rate may be speeding up or slowing down.

The remarkable predictability of Type Ia supernovae is what makes this project possible. Although all Type Ia supernovae don't have the same brightness, their absolute luminosity can be learned by watching how quickly each supernova fades away. Type Ia supernovae in nearby galaxies are so predictable that the time the supernova explosion began can be determined just from a look at its spectrum, and the most distant supernovae also turn out to have precisely the right spectrum on the right day of the explosion. "The real similarity of the details of these events," says Perlmutter, "can be seen in the beautiful spectra we get from the Keck Telescope in Hawaii, the largest in the world." Researchers breathed a sigh of relief when it was clear that Type Ia supernovae that exploded when the universe was half its present age behave essentially the same as supernovae do now, for this eliminated one worry about the reliability of the project's results—the question whether Type Ia supernovae have been different in different epochs.

Because the most distant supernova explosions appear so faint from Earth, happen at unpredictable times, and last for such a short while, the team performs a tightly choreographed sequence of observations, using telescopes around the world and the Hubble Space Telescope. Some team members survey distant galaxies using the largest telescope in the Andes Mountains of Chile, while others in

The dome of the Keck Telescope on Mauna Loa.
(Ernest O. Lawrence, Berkeley National Laboratory)

Berkeley, California, receive that data over the Internet and analyze it to find supernova candidates. Once they find likely supernovae, they rush out to Hawaii to confirm that these *are* supernovae and measure their redshifts. Team members at telescopes outside Tucson and on the Canary Islands are meanwhile standing by to measure the same supernovae as they fade. The Hubble Space Telescope is summoned into action to study the most remote of the supernovae, whose distances make them too difficult to measure accurately from the ground.

By January 1998, Perlmutter's team had analyzed forty of the roughly sixty-five supernovae so far discovered by the project. Only a little earlier they had reported that the cosmic expansion rate seemed to have slowed down very little, if at all. Now Perlmutter was ready to report that "all the indications from our observations of supernovae spanning a large range of distances are that we live in a universe that will expand forever. Apparently there isn't enough mass in the universe for its gravity to slow the expansion to a halt."

In March 1998, a second research group reported similar findings. This team was headed by Brian Schmidt of the Mount Stromlo and Siding Spring Observatory in Australia and included Adam Reiss, a young astronomer at the University of California at Berkeley, and Robert Kirshner from Harvard-Smithsonian. They reported that they had found indication that the expansion rate is approximately 15 percent greater now than it was when the universe was half its current age.

No sooner were the words out of Perlmutter's and Schmidt's mouths than speculation began in earnest about what this news might mean for inflation theory and for the cosmological constant. Inflation theory predicts a flat universe, but the new findings were indicating an open universe. Or were they? One extremely intriguing implication of the discovery was that these teams of astrophysicists might actually be looking at the first strong observational evidence that there is a repulsive force operating in the universe, that the universe is indeed getting an antigravity boost from somewhere. The evidence, said Perlmutter, strongly suggested a cosmological constant.

No one was jumping to the conclusion that there are no other possible explanations. Michael Turner reflected the caution of the scientific community when he said, "If it's true, this is a remarkable discovery. It means that most of the universe is influenced by an abundance of some weird form of energy whose force is repulsive." Schmidt said his own reaction was "somewhere between amazement and horror. Amazement, because I just did not expect this result, and horror in knowing that it will likely be disbelieved by a majority of astronomers who, like myself, are extremely skeptical of the unexpected." Reiss commented, "We are trying not to rush to judgment on the cosmological constant. There could be some other sneaky little effect we have overlooked, something that makes the supernovas dimmer and appear to be farther away than they really are, or some variation in the behavior of more distant supernovas that are deceiving us."

In spite of such reservations, it seems Schmidt had overestimated his colleagues' skepticism, for by May a straw vote at a workshop at Fermi National Laboratory indicated most scientists present agreed the two teams had made strong cases for an accelerating expansion rate and the existence of something resembling a cosmological constant.

It was all beginning to fit: The slowing down caused by the mass density of the universe appeared to be overwhelmed by the speeding up caused by the cosmological constant. Study of the relationship was telling researchers how much larger the energy density due to cosmological constant energy must be than the energy density due to mass density. The discovery also held out hope for solving the age of the universe glitch, allowing the universe to have expanded more slowly at an earlier age.

But is the secret ingredient really that old ghost, the cosmological constant? Some have been calling the enigmatic presence "X-matter" or "quintessence" (named after a fifth element suggested by Aristotle)—speculative concepts in which textures in the early universe created conditions for a cosmic background energy. By the time of the workshop at FermiLab, cosmologists were referring to the "missing energy" of the universe in the same way they had long spoken of the "missing matter." Some were calling it "funny energy." The mystery of what it is awaits early twenty-first century physicists.

The Supernova Cosmology Project and Brian Schmidt's group

hope to study supernova observations even farther back in time, to about ten billion light years distance. There are also proposals for studies involving new X-ray astronomy spacecraft and for surveys of the cosmic microwave background radiation from the ground and from space, with the Microwave Anisotropy Probe (MAP) in 2000, and, in 2004, with the Planck satellite. Clues to the density of the universe and the value of the cosmological constant are encoded in that radiation as the minuscule temperature variations that Smoot and his colleagues found.

A Theory Struggles to Cope

With these new discoveries of the late 1990s, inflationary Big Bang theorists found themselves torn between glee and discouragement. One of the theory's greatest assets, its ability to solve the flatness problem, had become a potential embarrassment, for researchers were continuing to find insufficient matter to maintain a flat universe. The evidence that the expansion rate was speeding up could be taken as another nail in the coffin of a flat universe. If the universe was open, what good was a theory that predicted a flat universe? The theory had gone to a great deal of effort to predict a situation that might simply not exist.

However, speculation is rampant about the cosmological constant value, and there is the possibility that this quintessence or funny energy might in fact make up the deficit left by insufficient mass density, producing precisely the omega-equals-1, flat universe that inflation theory predicts. Furthermore, since inflation theorists had already suggested that the cosmological constant might be the agent behind the inflation period in the early universe, observational evidence for its existence would be all to the good.

Another possibility for redeeming the theory is to reinterpret it to predict an open universe rather than a flat one, but most theorists balk at such a move. There is that danger—*indeed* reminiscent of Ptolemaic theory—of adding complications to a theory until it flounders because it can explain and predict too many contradictory findings.

This chapter has only barely sketched the problem of the elusive omega, given a small taste of the complications and the high hopes of modern research, and provided some background to understand announcements that inevitably will come in the next months and years. No one really knows at the moment whether the arguments will continue for a long time with more and more disparate voices and conflicting data or whether there might actually be definitive answers soon to questions that very few scientists have actually expected to see answered in their lifetimes.

CHAPTER

9

Lost Horizons

66 'Reality,' whatever that may mean. 99
Stephen Hawking

Whether there is an edge to the universe and what, if anything, might be beyond it are old questions. German astronomer Heinrich Wilhelm Olbers, who lived from 1758 to 1840, pointed out what is now known as Olbers's paradox: If space has no edge and is infinite and contains an infinite number of stars, the night sky should be as bright as the Sun. It isn't. Olbers wasn't the first to worry about that and certainly not the last. Suppose, instead, space is infinite, but the number of the stars is not, and the stars are limited to some sort of system "inside" infinite space. That creates another problem: The star system would collapse because of the stars' mutual gravitational attraction. Try to solve that by saying the star system rotates and its centrifugal force keeps it from collapsing, and someone will surely think to ask: In relation to what is it rotating, since it's the only thing in an infinite universe?

An expanding universe takes care of some of those problems while introducing other challenges, particularly to nonexpert thinkers: The balancing act summed up in the formula for omega and such proposals as Friedmann's first model in figure 6.1—a universe that is not infinite in size but nevertheless doesn't have any edges or boundaries in space (though it does in time), and the paradox of something that expands but doesn't expand "into anything."

Theorists at the cutting edge of physics and astrophysics try to help us get our minds around these counterintuitive concepts with stories of spheres and balloons, saddles and cones. That leaves us with another puzzle. How literally are any of these theoretical descriptions, metaphors, and analogies meant to be taken? The problem sixteenth- and seventeenth-century scholars had in deciding whether to accept Copernican theory as an ingenious and useful hypothesis or as literal truth was not a foolish one. Clearly, all theories are not equal when it comes to how much they must be taken as "reality." Many physicists think that inflation theory is a description of something that very possibly happened—either that or something like it. Fewer are prepared to give wormhole theories or Hawking and Jim Hartle's no-boundary universe that much credence, but they do not reject those or treat them as fantasy. Then there is Hawking's recent proposal that the universe sprang into being from nothing, in the form of a particle of space and time resembling an extremely small, slightly irregular, wrinkled sphere in four dimensions—the "pea instanton." Though he may be right, you aren't required to salute that yet. Are any of these theories likely to be confirmed from observational and experimental evidence in the way the Big Bang theory has been? Most physicists and astrophysicists would say that may happen for inflation theory. For the others, probably not, though no one entirely rules out the possibility.

When Time Is Space

In a universe where on the large scale everything is moving farther and farther apart, reversing the direction of time shows everything getting closer and closer together. In the late 1960s, Hawking and

Stephen Hawking.
(AP Wide World Photos)

Penrose, taking off from Penrose's earlier work on black holes, showed that, in Hawking's words, "if general relativity is correct, any reasonable model of the universe must start with a singularity"—that is, a point at which everything humans will ever be able to observe in the universe was compressed in a point of infinite density. A singularity is a dead end. All the theories of classical physics become useless there. No one can predict what would emerge from the singularity, and it's no use asking what happened before it. No wonder that when they find themselves locked out like this, physicists, whatever their religious or philosophical persuasion, don't take it with good grace.

Hawking was disinclined to let the Big Bang singularity lie, and he went on tinkering with his own ideas about what happens when things are very compressed, either in the center of a black hole or in the very early universe. Eventually he and American physicist Jim Hartle decided to employ a device called "imaginary time."

It isn't accurate to speak of Einstein's erasing the difference between the space dimensions and the time dimension. There remains a distinction in relativity theory between them. However, theorists have found a way around this distinction by making the time coordinate an "imaginary number" (see box). In that case there no longer are three dimensions of space and one of time, or four dimensions of space-time. Instead it's possible to think of them as four dimensions of *space*.

An **imaginary number** is a number that when squared yields a negative number. In everyday mathematics, the square of 4 is 16. The square of −4 is also 16. How can you square any number and arrive at −16? In the seventeenth century, Gottfried Leibniz invented imaginary numbers to provide an answer to that question. The square of imaginary 4 is −16. The square root of −16 is imaginary 4.

Hartle and Hawking's use of imaginary numbers and imaginary time is subtly different from the way others have employed them. They don't merely use this mathematical trick to solve a problem and then return to the more familiar concept of time. In their "no-boundary" model, imaginary time is something that actually shapes the universe. They point out that imaginary time is a device used in quantum theory, and that is the theory dealing with the very small—atoms and elementary particles. In the early universe, everything was that small. The principles and concepts of quantum mechanics could be expected to apply.

Hartle and Hawking's suggestion is that if you were able to travel back toward what everyone has been assuming was the "beginning" (the singularity), you would find, just short of reaching it, that (in imaginary time) it becomes meaningless to talk about "past" at all. In a situation where there are four space dimensions and no time dimension, chronological time—with its well-defined past, present, and future—would not exist, and with it would go all the vocabulary for describing chronological time. No more *yesterday*, or *always*, or *past*. Discussions about a *beginning* or *before the beginning* would have no meaning.

Hawking asks us to imagine traveling south on the face of the Earth. We can speak of ourselves as traveling south until we reach the South Pole, but there the concept of south is meaningless. No one asks what is south of the South Pole. This is also a good analogy because there is no edge or boundary or beginning at the South Pole. Similarly, there are no boundaries or edges or beginning in the Hartle-Hawking no-boundary universe—none in space and none in time. Does it follow that in this model time and space stretch to infinity? No. Just as the surface of the Earth is finite, space and time are finite in the no-boundary universe.

Hartle and Hawking refer to the no-boundary universe as a proposal, not a theory. There are no direct observational data to support it. It is a wild but not illogical leap of imagination. In the 1980s and 1990s, Hawking and others proceeded to ask what sort of universe would result from this no-boundary situation and to explore the model's connections with the observable universe of today. Needless to say, the calculations are extremely complex, and so far they've been carried out only in simple models. However, until the late 1990s, calculations seemed to indicate that there was no mathematical inconsistency between this proposal and the universe as we observe and experience it, or with well-accepted theories of modern physics. It doesn't compete with Big Bang theory, for in real time— the time in which we live—it would still appear that there was a singularity and a Big Bang at the beginning of the universe.

Now comes the new challenge of the late 1990s. The expansion rate seems to be speeding up. Could that happen in a no-boundary universe like Hartle and Hawking's, or does it rule out their model? An accelerating expansion rate might mean the universe is open and will go on expanding forever. A no-boundary universe is analogous (though in more dimensions) to the shape of a sphere—like the Earth—which means it *won't* go on expanding forever. It is a closed universe, one that contracts again to a "north pole."

But Hawking and Neil Turok, a colleague at Cambridge University, have thought of a way to look at the no-boundary universe as *either* spherical and closed and finite *or* open and infinite. Before, Hartle and Hawking had asked us to visualize the history and future of the universe as analogous to a sphere, and to imagine traveling south on the Earth as analogous to traveling into the past. Now they ask us to take that sphere and open out the top of it, the result being a shape something like the bell of a tuba. You can think of this tuba universe as having a "south pole" but no "north pole." In other words, no eventual collapse to a Big Crunch.

The Observable Universe Grows Tiny

In the late seventeenth century, measurements of the solar system caused the Earth to seem very small. In the nineteenth century, the solar system turned out to be tiny and lonely in the context of the distances to the nearest stars. In the twentieth, the Galaxy became miniscule in relation to the large-scale structure of the universe. Could it be that the universe itself is insignificant compared to something bigger still?

Inflation theory makes suggestions about how our universe may relate to a vastly larger picture. Recall the balloon analogy in chapter 6 that demonstrated how inflation theory solved some of the problems in Big Bang theory. (Inflate an imaginary balloon a little to represent the expansion of the universe before the inflationary period, pause to mark a tiny red dot on the surface of the balloon, and then inflate the balloon to a truly remarkable size. The tiny red dot itself becomes huge.) It was the red dot, not the balloon, that represented the entire observable universe. Our "universe" turned out to be possibly only a very small fraction of everything there is. If you drew a great number of red dots on the balloon, when the gravitational repulsive force came, what would happen to the other dots? Would any of them also expand? Would they become universes too? It is unlikely that anyone will ever be able to discover whether our universe is unique. If what we call the universe—this vast panoply of stars and galaxies that telescopes are revealing— started out as only an infinitesimal portion of what existed at the beginning, how can human beings ever hope to discover the ultimate structure and dimensions of *everything*?

For anyone hoping that the quest to measure the universe will culminate with the revelation of the dimensions of everything there is, such news will inevitably be disappointing. Big Bang theory has it that the observable universe amounts to 75 to 90 percent of the total universe. Inflationary Big Bang theory says that the observable universe is only a tiny fraction of the total, and no one knows what fraction. If there were an infinite number of dots on the balloon, even talking about a fraction is incorrect.

Andrei Linde has proposed something even more extensive: that each microscopic region that inflates is made up in turn of microscopic subregions, which inflate and are in turn made up of microscopic subregions—and so on and so forth—an eternal inflationary universe scheme. As Linde describes it, instead of being a single expanding fireball created in the Big Bang, "the universe is a huge, growing fractal. It consists of many inflating balls that produce new balls, which in turn produce more balls, ad infinitum."

A Labyrinth of Universes

Is there any way to travel from one balloon or dot or ball to another? The idea of wormholes isn't new, nor is the notion (much utilized in science fiction films and television series) that they might offer a way of traveling to distant regions and times in the universe or to other universes. Wormholes are not science fiction. They were "discovered" as a solution to Einstein's field equation in 1916 not long after he produced it. In the 1950s, John Archibald Wheeler led a research group that studied wormholes. Wheeler introduced the possibility of "quantum wormholes." It was these that captured the attention of Sidney Coleman of Harvard and Stephen Hawking in the 1980s. Coleman and Hawking took a particular interest in the possibility that such wormholes are part of the process in which new universes come into existence.

Quantum wormholes as these theorists propose them are extremely small, only about 10^{-33} centimeters across. (Written out as a fraction, that is 1 as the numerator and 1 followed by thirty-three zeros for the denominator.) These tiny holes flicker into existence and then vanish after an interval too short to imagine. Again, think of an enormous balloon, the cosmic balloon, our universe. Picture dots on the balloon's surface. This time they represent not fledgling universes but stars and galaxies. Einstein predicted that massive objects curve space-time, and the dots are doing that to the balloon's surface, causing tiny dimples and puckers. In spite of these, the surface is relatively smooth, even when examined through a microscope. It will take a more powerful microscope than any possible with present technology to reveal that it is not smooth after all. What Hawking and Hartle are picturing is something no human being has ever seen except in theory or imagination: what the universe looks like on a scale 1,000 million million million times smaller than the scale of an atomic nucleus. The surface at this magnification is vibrating furiously, creating a frothy foam. At this level, fluctuations in the curvature of space-time are not big, smooth curves like swells on the ocean. They are continuously changing ripples, crinkles, and swirls. The "surface" is hardly a surface at all. It is like a bubble bath.

There are those, including Hawking, who like to quote the dictum of quantum theory that what is not forbidden can and will occur. It isn't surprising, then, to hear him say that under high enough magnification, the quantum fluctuation becomes such that there's a probability we'll find it doing "anything." The cosmic balloon might develop a minuscule bulge in it, and that in turn could become another tiny balloon, attached to the parent balloon by a narrow neck. The neck is a wormhole; the balloon is a baby universe.

It hardly needs saying that there are no experimental or observational data to support this theory. Hawking is pessimistic about any tests revealing the existence of wormholes. No one expects to find direct observational evidence, for wormholes exist only in imaginary time, and even if that were not the case, their size rules out seeing them. However, the newborn universes attached to these umbilical cords need not continue to exist only in imaginary time or stay small. If the theory is correct, our own universe may have begun this way. A new universe might end up extending many billions of light-years. Of course, with one universe spawning many more universes, and those many more yet, there would be a never-ending labyrinth of them. Successfully measuring the dimensions of our own universe would offer no clue about what portion of the whole this represents.

These brief introductions to four late-twentieth-century models of the universe and beyond supply plenty of material for speculation, but the images described here do not allow you and me to say the universe *is* or even *may be* shaped like a tuba, a sphere, a pea, or a balloon—or is a dot on one balloon in the bunch being sold by the balloon man at some supercelestial carnival. These analogies are the closest theorists can come to representing in ordinary descriptive language what is only completely describable in the language of mathematics.

Einstein often spoke of the gift of fantasy being essential to the work he did. Certainly these intellectual descendants of his who speculate about how the universe fits into a larger context have that gift. In one sense they are all remarkably inventive yarn spinners. And yet their theories and proposals are not science fiction. This is fantasy tethered to the known world by hefty guy wires of mathematical equations. Nevertheless, no one yet knows whether these theories will be remembered merely as ingenious curiosities, like Kepler's linking the planetary orbits to the regular solids or musical phrases, or whether, like his three laws of planetary motion, they might turn out to be among the most significant advances in the history of astronomy. With present technology, it isn't even possible to make an educated guess which it will be.

Magnificent Enigma

66 Far and few, far and few
   Are the lands where the Jumblies live.
   Their heads are green, and their hands are blue,
   And they went to sea in a sieve. 99
   from "The Jumblies" by Edward Lear

In the spring of 1997, one of the most breathtaking views of the Hale-Bopp comet was from the Lofoten Islands off the coast of Norway, north of the Arctic Circle. Fog often shrouds this remote, jagged land and seascape. But that spring, when the mists cleared, the comet was clearly visible, suspended against a background of stars above an eerie billowing drapery of northern lights—all reflected in the water. Every now and then a streak of color shot into the sky, soared in an arc over the comet, and disappeared among the stars.

Over the comet? Among the stars? Spectators on the Lofoten Islands might have thought so had they not known that the stars were much farther away than the comet, not friendly sparkles but enormous, blazing infernos. The comet and its tail, which you could hide behind your thumb held at arm's length, was actually millions of miles of light, dwarfing the Earth. The northern lights are on our own doorstep by comparison and do not come anywhere near soaring in an arc "over the comet."

In these chapters, we've followed human beings as they've pondered the magnificent enigma of that same sky for thousands of years with wonder, loneliness, and, as Galileo put it, "intense longing." Men and women have written poems and sung about its beauty, worshiped it, found evidence there of a Creator God or no God at all, scrutinized it with telescopes, ventured a little way out. And, with remarkable success, they've measured the universe.

All tamed then? Is that how the story ends? The great mystery reduced to numbers and graphs?

One of my favorite analogies for the progress of science is one I learned from Hermann Bondi: Our scientific knowledge is an island—an island of "What We Know"—that lies in the midst of a vast sea of the Unknown. As long as human beings have lived on the Earth, they have been adding to that island, and the labor continues still at a frenzied pace. So the island of What We Know grows larger and larger, spreading in all directions. Sometimes a cliff crumbles back into the sea, or we forfeit acreage in a hurricane, or a tidal wave sweeps away a promontory . . . and someone cautions Mr. Elmendorf that a large portion of the land could shift alarmingly any day. But the island does grow inexorably, and it is astounding how much we now know! Unless the sea of the Unknown is infinite, it surely must be shrinking.

Perhaps. But we notice something strange happening. Every time we add to the island of What We Know, the coastline—the line where we run up against the Unknown—grows longer. There are more and more locations where we encounter what we *don't* know.

In this book, we've witnessed the living out of this parable over the course of more than two millennia. There has been evidence of it in the burgeoning cast of characters as the story progressed. In the early chapters, though these dealt with very long time spans, there were only a handful of men on the water's edge. Later chapters discussed much shorter periods of time, but the number of men and women involved grew. In chapters 7 and 8, it was barely possible to list all the names of those building onto the headlands, peering at the horizon, and putting out to sea both in small craft and exceedingly expensive contraptions.

The increase in sheer numbers is, of course, partly due to a growing world population and wider availability of university-level education. The progression, as it manifests itself in this book, is also somewhat explained by the fact that the more hindsight we have, the better we're able to discern which were the significant insights and discoveries. If I had been writing this book at the time of Copernicus, my discussion of "contemporary" astronomy would have highlighted many more researchers, books, and ideas of that day than I mentioned in chapter 2. Yet, four hundred years later it is easy to tell what work in the sixteenth century was leading to a dead end and what "cracked a door" to the future, and to see that Copernicus dwarfed his contemporaries.

But the expanding number of people and projects also bears witness to the fact that the coast of our island is growing. There is more room on the shoreline every day, and people are eager to fill it. We're aware of so many more questions than our ancestors thought to ask, more areas that need investigation, more trails of evidence to be followed, anomalous details to make sense of, complexity to be unraveled, paradoxes to get our stubbornly intuitive minds around. Ptolemy, Copernicus, Galileo had no idea how mysterious this universe would turn out to be!

If the mystery isn't lost, perhaps the simplicity and single-mindedness of earlier science is? Has the coastline become too long? Too many names, too many teams, too much specialization, too many directions to go? Has it become nothing else but the exponential accumulation of arcane knowledge—more than anyone can ever fit together in a coherent picture?

Nature itself has made certain it isn't like that. We have learned that this reaching into the Unknown, while it certainly encounters bewildering complications, always seems to grasp simplicity. The story appears to work backward: At the end of the Hellenistic era, Ptolemy's explanation of the heavens was as complicated as the mechanics of Disneyland. Copernicus's description was a move toward simplicity. Kepler's more so. Newton's understanding was even more concise. Einstein's, simpler yet. Observations like those Galileo made with his telescope and those we are making with the Hubble Telescope can be puzzling, can seem almost beyond possibility of explanation. Human genius searches for and often finds the beautiful symmetry that underlies and makes sense of the confusion.

I began this book by recalling how my father, my brother, and I measured a windmill. We did it in an old-fashioned way, with hindsight, if you will. We knew that there were ways to measure it more directly, that its height was a number that could be found out precisely. We were not pushing our mathematics and technology to the limits by any means. However, the men and women in this saga of cosmic measurement *have* been doing that and more.

It is a particular fascination of mine to try to put myself in the place of those in earlier centuries who *didn't know what was coming next*. When I try to do this with the characters in this book, I am immediately reminded that at every stage of history, it was impossible for them to anticipate that what they could not do, their descendants *would* be able to do. Perhaps that's why they were so often willing to go out on a limb, to build on shaky assumptions, to fashion a precarious ladder, to put their faith in ill-understood calibrators . . . because they had no way of knowing whether it ever would be possible to stand on firmer ground.

Whatever the motivations, we have been an impatient and irrepressible lot . . . measuring the parallax of Mars in full knowledge that the uncertainty of the measurement would imperil the results . . . scrambling up the scaffolding to a giant telescope before it was safely braced . . . battling to measure stellar parallax long before the technological moment arrived . . . grasping at Cepheids as handholds into the universe without knowing the distance to even one of them . . . devising a formula for omega while our understanding of what goes into the formula is still worse than vague. In each epoch, we have hoped that the impossible would later become easy. But we couldn't know, and we weren't willing to wait.

It might all have been less messy had we waited. It might in fact have more resembled what one would like to think the history of cosmic measurement has been. Many accounts give the impression that after Copernicus we "knew" the arrangement of the solar system. After Cassini we "had" the measurements to the planets and "were sure of" the length of the baseline afforded by Earth's orbit. In the 1950s, we finally "discovered" the size of the universe. It all moved by solid increments.

It hasn't been like that. We have groped, guessed, doubted one another, made missteps, built the rungs of the ladder too close together, felt it buckle beneath our feet, fought for a hold. This book ends not on the coastline of the island of What We Know but on jetties built far out from shore . . . even on small, frail crafts almost out of sight of land . . . at sea in a sieve. But that is nothing new. Indeed that is precisely where we have been in every chapter of this book, at every stage of this history.

The nature we've striven to understand has fought back by showing us our place. Sometimes that's been a severe comeuppance, but it isn't all bad news. To be sure, human beings are not the center of the universe, and they are not large by universal standards, but they aren't small either. Not the largest nor the smallest things around by a long shot.

There's another way to measure us. We are the most complex thing we have yet discovered in the universe. The human mind is still largely unexplained. The human situation is unfathomable. How paradoxical that with motives and longings and limitations rooted in the confusion of who we are, we probe the depths and heights, often with complex mathematics as our only tool . . . on a quest to discover not more complication, but simplicity!

"Who hath stretched a measuring line across it?" God taunts Job in the Scriptures. Shall we raise a timid hand and venture, "I think . . . well . . . actually . . . *we* have"? Maybe. Maybe not. For it is still a great mystery how large our island is—this treasured, hard-won, incalculably valuable, perhaps tiny island of human knowledge—compared with the sea.

These notes do not provide a comprehensive bibliography, nor do they include all the sources used for this book. They are intended as a guide for readers who would like to explore further. For the most part, the titles listed here are on about the same level of technical difficulty as this book.

There are no books that trace the lengthy history of cosmic measurement that is covered in this book, but for a fine general survey of the history of astronomy, look for John North, *The Norton History of Astronomy and Cosmology* (Norton, 1995); for a comprehensive history of the telescope, see Henry C. King, *The History of the Telescope* (Dover, 1979).

Three sources awakened my initial interest in the subject of cosmic measurement: James Cornell, ed., *Bubbles, Voids, and Bumps in Time: The New Cosmology* (Cambridge University Press, 1989), containing articles on discovering, measuring, mapping, and weighing the universe by Alan P. Lightman, Robert Kirshner, Margaret Geller, Vera C. Rubin, Alan Guth, and James Gunn; Harlow Shapley's account of his research, "Measuring the Universe," reprinted in Timothy Ferris, ed., *The World Treasury of Physics, Astronomy, and Mathematics* (Little, Brown, 1991); and Owen Gingerich's "Let There Be Light," reprinted in the same book.

1. A Sphere with a View, 400–100 B.C.

On scientific thought and achievements in antiquity, see Marshall Clagett, *Greek Science in Antiquity* (Abelard-Schuman, 1955); David C. Lindberg, *The Beginnings of Western Science* (University of Chicago Press, 1992); Geoffrey E. R. Lloyd, *Greek Sciences after Aristotle* (Norton, 1973); Otto Neugebauer, *The Exact Sciences in Antiquity* (Dover, 1969); Otto Neugebauer, *A History of Ancient Mathematical Astronomy*, 3 vols. (Springer-Verlag, 1975); and George Sarton, *A History of Science*, vol 2, *Hellenistic Science and Culture in the Last Three Centuries B.C.* (Harvard, 1959).

The writings of some of the ancient scholars are collected in M. M. Austin, ed., *The Hellenistic World from Alexander to the Roman Conquest: A Selection of Ancient Sources in Translation* (Cambridge University Press, 1981).

Thomas Kuhn's classic account, which includes many of the people and concepts in this chapter and those to follow, is *The Copernican Revolution: Planetary Astronomy in the Development of Western Thought* (originally printed by Harvard

University Press, 1957). Kuhn's technical appendix explains Eratosthenes's measurement of the circumference of the Earth and Aristarchus's measurements having to do with the Sun and the Moon. My own understanding of the latter was also helped by Albert van Helden, *Measuring the Universe: Cosmic Dimensions from Aristarchus to Halley* (University of Chicago Press, 1985) and Rocky Kolb's explanation in *Blind Watchers of the Sky* (Addison-Wesley, 1996).

2. Heavenly Revolutions, 100–1600 A.D.

Owen Gingerich tells of Mr. Elmendorf's offer in his article "Let There Be Light," which is reprinted in Timothy Ferris's book cited at the beginning of the note section. The article originally appeared in Roland Mushat Frye, ed., *Is God a Creationist?* (Scribner's, 1983)

A fine translation of Ptolemy's *Almagest* is by G. J. Toomer (Springer-Verlag, 1985). For Ptolemaic astronomy in the Middle Ages, Olaf Pedersen and M. Pihl, *Early Physics and Astronomy* (Cambridge University Press, 1993; reprint of 1974 edition). To learn more about medieval Islamic astronomy, read David A. King and George Saliba, eds., *From Deferent to Equant: A Volume of Studies in the History of Science in the Ancient and Medieval Near East in Honor of E. S. Kennedy* (New York Academy of Sciences, 1987).

The most complete edition of Copernicus's writings comes from the Polish Academy of Sciences. The first volume reproduces Copernicus's manuscript of *De Revolutionibus* (Warsaw and Cracow, 1973). The series also includes English translations. Facsimiles of *De Revolutionibus* are also to be found on the Web. The introduction to Noel M. Swerdlow and Otto Neugebauer, *Mathematical Astronomy in Copernicus's De Revolutionibus* (Springer-Verlag, 1984) gives information about the life of Copernicus. The book analyzes the mathematics of Copernicus's work. Owen Gingerich, *The Great Copernicus Chase* (Cambridge University Press, 1992) is a fascinating collection of articles written for *Sky and Telescope* magazine, several of which deal with Copernicus. Patrick Moore, *The Great Astronomical Revolution* (Albion, 1994) has chapters on Copernicus. Fred Hoyle does a masterful job, in his *Nicolaus Copernicus: An Essay on His Life and Work* (Harper and Row, 1973), of erasing the notion that Ptolemaic astronomy was "wrong." Kuhn (see reference for chapter 1) is an invaluable source on Copernicus.

3. Dressing Up the Naked Eye, 1564–1642

Two biographies of Johannes Kepler are: Max Caspar, *Johannes Kepler* (Dover, 1993); Arthur Koestler, *The Watershed* (University Press of America, 1985). For more details on Kepler's work, see Bruce Stephenson, *The Music of the Heavens* (Princeton, 1994). For translations of some of the books by Kepler discussed in this chapter, see *Mysterium Cosmographicum: The Secret of the Universe*, trans. A. M.

Duncan (Abaris Books, 1981); William H. Donahue, *Johannes Kepler: New Astronomy* (Cambridge University Press, 1992).

There are many biographies of Galileo. One classic is Stillman Drake, *Galileo* (Oxford University Press, 1980). Two recent ones are Michael Sharratt, *Galileo: Decisive Innovator* (Blackwell, 1994); and James Reston, Jr., *Galileo: A Life* (Harper-Collins, 1994). Drake's *Galileo at Work: His Scientific Biography* (University of Chicago Press, 1978) highlights how crucial was the distinction between Copernicus's astronomy being a useful mathematical hypothesis and representing physical reality. Also from Stillman Drake comes a collection of Galileo's writings (abridged): *Discoveries and Opinions of Galileo* (Anchor, 1990); and a translation of *Dialogo: Dialogue Concerning the Two Chief World Systems* (University of California Press, 1967). Galileo was an entertaining, clear writer, and his book is fun to read.

For those interested in getting behind the legend of Galileo's relationship with the Catholic Church and his trial; Maurice A. Finocchiaro, *Galileo Affair: A Documentary History* (University of California Press, 1989); Rivka Feldhay, *Galileo and the Church: Political Inquisition or Critical Dialogue* (Cambridge University Press, 1995); and three wonderfully evenhanded shorter retellings: Kolb (see reference for chapter 1); Gingerich (see his referenced *The Great Copernicus Chase* for chapter 2); and John Hedley Brooke, *Science and Religion: Some Historical Perspectives* (Cambridge University Press, 1991).

Galileo's study of the phases of Venus is well described in a chapter of Gingerich's book (see reference for chapter 2).

4. An Orbit with a View, 1630–1900

The description of parallax measurement and the diagrams explaining it were developed with the help of Barbara Quinn, P. Susie Maloney, and David Vetter.

For Newton's measurement of stellar distances, discussions of Halley, Molyneux, and Bradley, and other material on this period, see Martin Harwit, *Cosmic Discovery: The Search, Scope, and Heritage of Astronomy* (Basic Books, 1981).

van Helden (see reference for chapter 1) goes into greater and more technical detail on the work of Gassendi, Cassini, Flamsteed, and Halley.

The classic biography of Newton is Richard Westfall, *Never at Rest: A Biography of Isaac Newton* (Cambridge University Press, 1990), and it has been abridged to a more manageable length as *The Life of Isaac Newton* (Cambridge University Press, 1993).

Henry C. King, *The History of the Telescope* (Dover, 1955) is a nontechnical, well-illustrated, comprehensive account, with biographical material on innovators such as Herschel and Fraunhofer.

Some of the anecdotes about the expeditions and attempts to study the transits of Venus in 1761 and 1769 are recounted in John North's book cited at the beginning of the note section.

Michael J. Crowe, *Modern Theories of the Universe* (Dover, 1994) gives a readable account of the history of spectroscopy and its part in building the cosmic distance ladder.

Another good source on this period is Milton K. Munitz, ed., *Theories of the Universe* (Free Press, 1957).

5. Upscale Architecture, 1750–1958

For more information on William Herschel, see J. L. E. Dreyer, *A Short Account of Sir William Herschel's Life and Work* (1912); and Gingerich, "The 1784 Autobiography of William Herschel" (in his book referenced for chapter 2). Dreyer printed edited versions of Herschel's collected papers with other biographical material in *The Scientific Papers of Sir William Herschel*, 2 vols. (Royal Astronomical Society, 1912).

For Lord Rosse's work, refer to Charles Parsons, ed., *The Scientific Papers of William Parsons, Third Earl of Rosse, 1800–1867* (London, 1926); and Patrick Moore, *The Astronomy of Birr Castle* (Mitchell Beazley, 1971).

An early history of the Harvard College Observatory is Solon Bailey, *The History and Work of Harvard Observatory, 1839–1927* (Harvard University Press, 1931).

For more on the Shapley-Curtis debate, read Robert W. Smith, *The Expanding Universe: Astronomy's "Great Debate"* (Cambridge University Press, 1982).

Crowe (see reference for chapter 4) includes excerpts from original papers by Herschel, Shapley, Curtis, and Hubble.

6. Coming Apart in All Directions, 1929–92

J. D. North, *The Measure of the Universe: A History of Modern Cosmology* (Oxford University Press, 1965; Dover, 1990) is a comprehensive history of cosmology in the first half of the twentieth century.

The classic biography of Einstein is Abraham Pais, *Subtle Is the Lord—the Science and Life of Albert Einstein* (Oxford University Press, 1982)—a wonderful book, though not an easy read.

For more biographical details on Edwin Hubble, see Gale E. Christianson, *Edwin Hubble* (Farrar, Straus, Giroux, 1995).

Jeremy Bernstein and Gerald Feinberg, in *Cosmological Constant* (Columbia University Press, 1986), reprint and annotate some of the original articles of Hubble, Einstein, Lemaître, and Friedmann.

For the story of Nikolai Kozyrev, see "Astronomers and Joseph Stalin," in Patrick Moore, *Fireside Astronomy* (John Wiley, 1992). This book contains anecdotes about many characters and events in the history of astronomy.

For an understanding of the expanding universe, see John D. Barrow, *The Left*

Hand of Creation: The Origin and Evolution of the Expanding Universe (Oxford University Press, 1994).

The story of the Big Bang theory has been told in many books. Two of the best for the general reader are John Gribbin, *In Search of the Big Bang.* (Corgi, 1987); and Robert Jastrow, *God and the Astronomers* 2nd ed. (Norton, 1992).

For the first decades of radio astronomy, see W. T. Sullivan III, *The Early Years of Radio Astronomy* (Cambridge University Press, 1984).

Alan E. Nourse's *Radio Astronomy* (Franklin Watts, 1989) is a splendid, simply written history of radio astronomy.

For the discovery of the cosmic microwave background radiation, see Robert Wilson's essay in B. Bertotti, R. Balbinot, S. Bergia, and A. Messina, *Modern Cosmology in Retrospective* (Cambridge University Press, 1990).

George Smoot describes the discovery of the "wrinkles" in that radiation in his book with Keay Davidson, *Wrinkles in Time* (Morrow, 1993).

In my book *The Fire in the Equations: Science, Religion, and the Search for God* (Eerdmans, 1994), I discuss religious belief and the Big Bang universe.

Alan Guth has written about inflation theory in *The Inflationary Universe* (Addison-Wesley, 1997).

For Stephen Hawking's theories and their implications, see Stephen Hawking, *A Brief History of Time: From the Big Bang to Black Holes* (Bantam, 1988); and Kitty Ferguson, *Stephen Hawking: Quest for a Theory of Everything* (Franklin Watts/Bantam, 1991).

Some of Roger Penrose's extraordinary ideas are presented in his book *The Emperor's New Mind* (Oxford University Press, 1989).

7. Deciphering Ancient Light, 1946–99

For a thorough and beautifully illustrated account of discoveries within the Galaxy, see Nigel Henbest and Heather Couper, *The Guide to the Galaxy* (Cambridge University Press, 1994).

Frederic Golden profiles Allan Sandage in "Astronomy's Feisty Old Man," in *Astronomy Magazine*, December 1997.

Malcolm Longair, *Our Evolving Universe* (Cambridge University Press, 1996) is a spectacularly illustrated account of current astronomy that also traces the history of the concept of a "grand design" and discusses sizes and distances in the universe.

Margaret Geller writes about the Geller-Huchra Wedge and related measurements of the large-scale structure of the universe in her article "Mapping the Universe: Slices and bubbles," in James Cornell's book cited at the beginning of the note section.

For information on the Hubble Deep Field, the Web site is: http://www.stsci.edu/ftp/science/hdf/hdf.html

For the Sloan Digital Sky Survey, see the Web site http://www.sdss.org

8. The Quest for Omega, 1930–99

Much of the material for this chapter comes from conversations with the scientists involved, from newspaper reports, from their Web sites, and from some of their papers.

Three recent books offer nontechnical discussions of these current issues: Martin Rees, *Before the Beginning* (Addison-Wesley, 1997); Timothy Ferris, *The Whole Shebang: A State of the Universe Report* (Simon and Schuster, 1997); and Longair (see reference for chapter 7).

Wendy Freedman's 1998 *Scientific American* article "The Expansion Rate and Size of the Universe" can be found at Web site http://www.sciam.com/1998/0398cosmos/0398freedman.html

For information about the computer simulations being carried out by Carlos Frenk's team, the Web site is http://star-www.dur.ac.uk/~frazerp/virgo/virgo.html

For Saul Perlmutter's group: www-supernova.lbl.gov

For Brian Schmidt's group: http://msowww.anu.edu.au

The most recent news is best gleaned from reports in *Nature* magazine, newspaper accounts such as those in the *New York Times Science Section*, and Web sites of the projects.

9. Lost Horizons

For more on Jim Hartle and Stephen Hawking's no-boundary proposal, see Hawking's and Ferguson's books referenced for chapter 6.

On inflation theory, see Guth's book referenced for chapter 6 and also Ferguson, *The Fire in the Equations*, referenced for chapter 6.

Andrei Linde explains his proposal in his article "The Self-Reproducing Inflationary Universe," in *Scientific American*, November 1994.

For more on wormhole theory, see Kip Thorne, *Black Holes and Time Warps: Einstein's Outrageous Legacy* (Norton, 1994); and Kitty Ferguson, *Prisons of Light* (Cambridge University Press, 1996).

Also see John D. Barrow, *The Origin of the Universe* (Basic Books, 1994).

Epilogue. Magnificent Enigma

The description of the appearance of the Hale-Bopp comet from the Lofoten Islands comes from Judy Anderson.

absolute magnitude: How bright a star looks from a distance of ten parsecs (32.6 light-years). More technically: The amount of light received from a star that is ten parsecs away. The absolute magnitude of a star doesn't change with distance. Just as a 100-watt lightbulb is still a 100-watt lightbulb no matter how much its brightness *appears* to change with distance, a star's absolute magnitude remains the same no matter how much its *apparent* magnitude changes with distance.

absolute zero: The lowest possible temperature, at which a substance contains no heat energy.

absorption lines: Dark lines in a spectrum produced when light from a distant source passes through cooler gas closer to the observer.

acceleration: The rate at which the speed of an object is changing.

action at a distance: The phenomenon of an object exerting a force on a second object across empty space, without the intervention of anything physical.

aether: Aristotle's fifth element, of which he thought stars and planets were made.

angular size: Astronomers describe the apparent size of an object in the sky in terms of its angular size. For example, if two lines are drawn from an observer on the Earth to the opposite edges of the Moon, the angle formed by the two lines where they meet at the observer is about $1/2$ degree. Another way of putting that is to say that the Moon "subtends" an angle of $1/2$ degree; or that it has an angular size of $1/2$ degree. The angular size of an object can't be converted into its true physical size unless the distance to the object is known. In calculating angular size: A circle has 360 degrees; each degree is divided into 60 minutes of arc or arcminutes; each minute of arc is divided into 60 seconds of arc or arcseconds.

apparent magnitude: How bright a star looks as viewed from the Earth. More technically: The amount of light received from a star as observed from the Earth. The apparent magnitude of a star changes with distance.

apparent size: As opposed to true size, apparent size is the size of a heavenly body as viewed from the Earth. The apparent sizes of the Sun and Moon are the same; their true sizes are not.

arcminute: See angular size.

arcsecond: See angular size.

astronomical telescope: A telescope based on an optical system first described by Kepler that uses only convex lenses. Kepler's telescope inverted the image, but after that problem was worked out, the astronomical telescope replaced the so-called Dutch telescope (the sort Galileo had used) in most serious astronomical work by the mid–seventeenth century. It provided a much larger field of view at equal magnifications.

atom: A unit of ordinary matter. The center of the atom is the nucleus, made up of protons and neutrons. Electrons orbit the nucleus.

bender: A massive body or galaxy or cluster of galaxies that is responsible for the bending of paths of light passing near it.

Big Bang: The state of enormous heat and density in which the universe probably began, and from which the universe has expanded and cooled to its present state; not necessarily a singularity.

Big Crunch: The collapsed state in which the universe might end.

binary star (or binary system): Double star system in which the two stars are bound together gravitationally and orbit their common center of mass.

black hole: The classical definition is a region of space-time from which nothing can escape unless it can travel at a speed greater than the speed of light.

blueshift: Displacement of the spectral lines in light coming from distant stars and galaxies that are moving toward Earth.

brightness fluctuation method: Technique used to calculate the distance to galaxies by measuring the unevenness in the brightness of the surface of the central bulge or near the center.

Cepheid variable: Pulsating variable star whose period of brightness variation is directly related to its absolute magnitude.

Chandrasekhar limit: About 1.4 times the mass of the Sun. The maximum possible mass of a stable cold star, above which it must collapse.

Charge-coupled device (CCD): A digital imaging system.

closed universe: Cosmological model in which the universe eventually stops expanding and collapses. A universe in which omega equals more than 1.

Copernican astronomy: The Sun-centered astronomy introduced by Nicolaus Copernicus.

cosmic distance ladder: Set of overlapping distance-measurement techniques by which astronomers over the centuries have bootstrapped their way to measuring farther and farther distances in the universe.

cosmic microwave background radiation: Radiation detectable in the microwave range of the spectrum (microwaves are radio waves with a wavelength of a few centimeters), composed of photons released from primordial cosmic material as it thinned out owing to the expansion of the universe, at the epoch of photon decoupling, about 300,000 years after the Big Bang. Discovered by Penzias and Wilson in the mid-1960s.

cosmological constant: The mathematical device Einstein added to his equations to allow the universe to remain static.

cosmology: The study of the origin, structure, and evolution of the universe. The word usually implies study of the universe on the large scale.

cosmos: A word that has become synonymous with "universe on the large scale," though it traditionally implies an orderly, harmonious universe.

critical density: The density that would allow the universe to balance eternally on the knife edge between expansion and collapse, expanding at precisely the right rate to avoid recollapse. Omega for a critical density universe equals 1.

curvature of space-time: Einstein's general theory of relativity explains the force of gravity as the way the distribution of mass in space-time causes something that resembles the warping, denting, and dimpling in an elastic surface by balls of different weights and sizes lying on it.

dark matter: Matter that can't be observed to radiate energy in any part of the electromagnetic spectrum.

deceleration parameter: Quantity designating the rate at which the expansion of the universe is slowing down.

decoupling: In this book, the separation of photons from particles of matter that made possible the cosmic microwave background radiation.

deferent: One of the devices used in Ptolemaic astronomy. It is a circle that carries another circle. The second circle is called an epicycle.

Doppler shift: A change in the apparent wavelength of radiation (such as light or sound), emitted by a moving body.

dynamics: The area of science that deals with how bodies respond and move as the result of the action of forces.

eccentric: In Ptolemaic astronomy, a circle on which a planet moves that is centered near but not precisely on the Earth.

electromagnetic spectrum: The complete range of electromagnetic radiation. Visible light is one form of electromagnetic radiation but represents only a very small part of the complete electromagnetic spectrum, which also includes gamma rays, X rays, ultraviolet rays, infrared rays, microwaves, and radio waves.

emission spectrum: A spectrum produced by an incandescent gas, consisting of only a few isolated colors. Each kind of gas has its own pattern.

epicycle: In Ptolemaic astronomy, a circle in which a planet moves, which in turn moves on a larger circle, the deferent, which is centered on the Earth.

equant: In Ptolemaic astronomy, a location interior to the deferent from whose point of view the angular velocity of the epicycle traveling on the deferent is constant.

flat universe: Cosmological model in which the universe expands at precisely the right rate to avoid recollapse. A universe in which omega equals 1.

flatness problem: The puzzle of why the universe appears to be balanced between being an open or a closed universe.

fractal: An object or situation that has the characteristic feature of self-similarity—an unending series of motifs within motifs repeated on every scale. The Mandelbrot set is the most familiar example of the fractal quality.

Galactic disk: If we picture the Galaxy as a giant fried egg, the disk is the white of the egg. This is the region in which the spiral arms are found.

Galactic halo: An aggregation of stars, globular star clusters, and thin gas clouds, centered on the core of the Galaxy and extending beyond the known borders of the Galactic disk.

gamma rays: Electromagnetic radiation of very short wavelength and very high energy.

globular clusters: Dense, jewel-like, spherical clusters of stars.

gnomon: A protrusion like the raised part of a sundial.

gravitational lensing: Multiple images and other distortions in the light from quasars and other distant objects produced by the warping of space-time by objects such as galaxy clusters in the foreground.

gravitational redshift: The shift in wavelengths toward the red end of the spectrum caused by a gravitational field.

gravity: One of the four basic forces of nature. Gravity always attracts, and it works on all levels, from the tiniest fundamental particles to the largest objects in the universe.

Heisenberg uncertainty principle: The principle in quantum physics according to which the position and momentum of a particle can never both be known with exactitude at the same time. It implies that empty space can never really be empty, but that instead there are always and everywhere in the universe tiny energy fluctuations.

horizon problem: The puzzle having to do with how few of the particles in the early universe would have had time to be in contact with one another as cosmic expansion began.

Hubble constant: The constant that denotes the expansion rate of the universe.

Hubble flow: Movement of objects in the universe that is directly attributable to the expansion of the universe.

imaginary numbers: Numbers that when squared yield a negative number.

imaginary time: Time measured using imaginary numbers.

impossible object: An object that can't really exist because it contradicts itself.

inflation theory (inflationary Big Bang theory): A theory that has the universe going through a very brief phase of extremely rapid expansion early in its existence.

inverse square: Technically speaking, "proportional to the power −2." For example, if you have two lightbulbs of equal wattage and place lightbulb B twice as far away from you as lightbulb A, lightbulb B will look only a fourth as bright as lightbulb A. The strength of the gravitational force also varies as the inverse square of the distance.

ionized: The state of an atom in which it has fewer or more electrons than is normal.

isotropy: The quality of being the same in all directions.

Kepler's laws: Three laws discovered by Johannes Kepler. The first states that a planet travels in an elliptical orbit around the Sun, with the Sun as one of the foci of the ellipse. The second states that the orbital velocity of a planet is directly related to where the planet is in its orbit, and the nearer the Sun the greater the velocity. The third shows the relationship between the lengths of time the planets take to complete their orbits and their distances from the Sun.

least squares: A statistical method of estimating values from a set of observations.

light-year: The distance traveled by light in one year in a vacuum. Equal to about 5.88 trillion miles or 9.4607 kilometers.

Local Group: The group of galaxies of which the Milky Way Galaxy is a member.

mass: How much matter there is in a body, or how much a body resists any change in its speed or direction.

megaparsec: One million parsecs. About 3.26 million light-years.

micrometer: An instrument for measuring very small dimensions. Placed inside a telescope, and having cross-wires that can be moved across the image with the turning of a screw, it serves as a ruler to measure the apparent size of the object being viewed.

moving cluster method: A technique that studies the way paths of stars in a cluster seem to converge or diverge, as an aid to measuring the distance of the cluster.

nebulae: Formerly described as distant, hazy, fuzzy celestial objects. Some are now known to be galaxies, whereas others are clouds of gas, dust, or debris from events such as supernovae.

neutrinos: Particles with very small mass that very rarely interact with any kind of matter and are extremely difficult to detect. First suggested in 1930 by Wolfgang Pauli as a way of explaining a loss of energy in some nuclear reactions.

nova: As modern astronomy since the 1930s makes the distinction, a nova is a less drastic flare-up of a star—usually a white dwarf star that is part of a binary system—as opposed to a supernova, which is an explosion that demolishes the entire star once and for all. Earlier astronomy used the term *nova* for any sudden appearance (flare-up) of a star where none had been observed before.

O and B stars: Massive, brilliant, short-lived stars, less than about 10 million years old.

omega: The mass density of the universe.

open universe: Cosmological model in which the universe expands forever, thinning out eternally. A universe in which omega is less than 1.

opposition: A planet is said to be in opposition when it is on the opposite side of the Earth from the Sun.

optical astronomy: Study of the universe by observing radiation in the optical (visible) range of the spectrum.

parallax shift: The apparent change in an object's position against the background when viewed from two different places.

parsec: A unit of distance equaling 3.26 light-years; a little more than 30 trillion kilometers.

pencil beam survey: A survey that is limited to a tiny circular area of the sky. The small circle under observation becomes a cone as it goes deeper into space, and the survey produces a cone-shaped three-dimensional map that keeps getting larger as it reaches farther distances.

perspicillum: An early name for the telescope.

photon: The "messenger particle" that carries the electromagnetic force. The photon is the particle of light and all other forms of electromagnetic radiation (gamma rays, X rays, radio waves, etc.).

primeval atom: Abbé Georges-Henri Lemaître theorized that there was a time when everything that makes up the present universe was compressed into a space only about thirty times the size of our Sun. He called that the "primeval atom."

proper motion: Change in position of a star, relative to other stars, over a period of years. More generally, the individual drifting of stars through space.

Ptolemaic astronomy: The combination, recombination, and reworking of Ptolemy's ideas as they were used by astronomers over the centuries. Earth-centered astronomy.

quantum physics (quantum mechanics): The study of the universe on the level of the very small—molecules, atoms, and elementary particles.

quasars: Pointlike sources of light whose redshifts show that they are billions of light-years distant. Thought to be the nuclei of young galaxies.

radio astronomy: Study of the universe by observing radiation at radio wavelengths.

radius: Shortest distance from the center of a circle or sphere to the circumference or surface.

redshift: Displacement of the spectral lines in light coming from distant stars and galaxies that are moving away from Earth. Redshift can also be caused by a gravitational field.

reflecting telescope (or reflector): A telescope in which incoming light is reflected from a primary mirror back to an eyepiece.

refracting telescope (or refractor): A telescope in which light is redirected (refracted) by a main lens to a specific point from where a second lens presents the image to the eye.

refraction of light problem: The way the Earth's atmosphere bends and smears light rays passing through it.

relativity, general theory of: Einstein's theory in which the gravitational force is explained as a curvature of space-time caused by the presence of mass, and in which the curvature of space-time dictates the movement of mass. As John Archibald Wheeler expresses it, "Spacetime grips mass, telling it how to move; and mass grips spacetime, telling it how to curve."

retrograde motion (or retrogression): When a planet seems to stop in its eastward motion and travel westward for a while.

singularity: The dimensionless point that may exist at the center of a black hole, where all the mass of the collapsing star has shrunk to infinite density and infinite space-time curvature. Singularity may also be used to name any point in space-time where the curvature of space-time becomes infinite, such as the Big Bang singularity.

solar mass: The mass of the Sun.

space-time: The combination of the dimensions of space and time. The space-time we experience has three dimensions of space and one dimension of time.

spectral lines: Bright and dark lines seen in the spectra of stars and other luminous objects.

spectroscope: An instrument for observing and studying spectra.

speed of light: The speed of light in a vacuum is approximately 186,300 miles or 299,800 kilometers per second, often rounded off to 186,000 miles or 300,000 kilometers per second.

static universe: A universe that doesn't expand or contract.

statistical parallax: A technique that assumes that the average velocity along our line of sight of all stars in a group is also the average velocity of them across our line of sight.

Steady State theory: A theory that the expanding universe was never in a state of much higher density than it is today—that there was no Big Bang. The theory hypothesizes that matter is constantly being created out of empty space.

stellar parallax motion (or annual stellar parallax): The apparent change in the positions of stars as seen from different parts of the Earth's orbit around the Sun.

sunspot: Darker patches that appear periodically on the face of the Sun.

supernova: The explosion of a star that completely demolishes it, as opposed to a nova, which is a less drastic flare-up.

transit: In this book, the passage of Mercury or Venus between Earth and the Sun in such a way that the planet shows up, to observers on Earth, as a spot traveling across the face of the Sun.

Tropic: Either of two corresponding parallels of latitude on the terrestrial globe. The Tropic of Cancer is about 23½ degrees north of the equator. The Tropic of Capricorn is about 23½ degrees south of the equator. These parallels of latitude in each case separate the torrid from the temperate zone.

Tully-Fisher method (Tully-Fisher relation): A method that uses a relationship between the luminosity and spectral line widths of galaxies to estimate their distances.

twenty-one centimeter line width: Hydrogen atoms, of which most of the interstellar matter spread throughout a spiral galaxy consists, emit radio noise at the wavelength of twenty-one centimeters. How much that spectral line is blurred by the rotation of the galaxy is directly related to the speed at which the galaxy rotates, and that speed is related to the galaxy's absolute magnitude.

Tychonic system: Tycho Brahe's proposed arrangement, with the Sun orbiting the Earth and all the other planets orbiting the Sun.

variable star: Star that changes in brightness periodically.

visible spectrum: The part of the electromagnetic spectrum in which the wavelengths are the right length for our eyes to receive.

wavelength: For a wave, the distance between two adjacent troughs or two adjacent crests.

wormholes: Theoretical connections between universes or between different places and times in the same universe.

INDEX